Microtechnology and Thermal Problems in Electronics

(MicroTherm 2013)

Journal of Physics: Conference Series Volume 494

Lodz, Poland
25-28 June 2013

Editors:

Ewa Raj **Zbigniew Lisik**

ISBN: 978-1-63266-719-9
ISSN: 1742-6588

Printed from e-media with permission by:

Curran Associates, Inc.
57 Morehouse Lane
Red Hook, NY 12571

proceedings.com

Some format issues inherent in the e-media version may also appear in this print version.

Printed by Curran Associates, Inc. (2014)

Institute of Physics
Dirac House, Temple Back
Bristol BS1 6BE UK

Phone: 44 1 17 929 7481
Fax: 44 1 17 920 0979

techtracking@iop.org

Additional copies of this publication are available from:

Curran Associates, Inc.
57 Morehouse Lane
Red Hook, NY 12571 USA
Phone: 845-758-0400
Fax: 845-758-2634
Email: curran@proceedings.com
Web: www.proceedings.com

TABLE OF CONTENTS

PAPERS

Preface

MicroTherm is an International Conference on Microtechnology and Thermal Problems in Electronics organised as a cyclic event since 1996. The success of the first seminar, which was devoted mainly to thermal management aspects, and the successive conferences have led us to the tenth edition. Since the first meeting, the scope of the conference has expanded, following the progress of electronics. Now, it covers subjects connected with extreme temperature, electronics, sensors and measurement techniques, modelling, simulation, wide band-gap materials, packaging and reliability, renewable energy sources and photonics with special emphasis on microelectronic technologies.

MicroTherm'2013 was held in Lodz, Poland, on 25–28 June 2013. The programme consistied of invited talks and nine regular sessions in the form of planar discussions and poster presentations, including a Students' Session. The Students' session gave an opportunity for students and young researchers to present their first achievements in the field of science.

The next MicroTherm Conference is going to be held on 22–25 June 2015, in Lodz — a beautiful, post-industrial city located in the centre of Poland. Please, feel invited to MicroTherm'2015 (www.microtherm.dsod.pl).

Ewa Raj and Zbigniew Lisik
Editors

Journal of Physics: Conference Series **494** (2014) 011002 doi:10.1088/1742-6596/494/1/011002

Peer review statement

All papers published in this volume of *Journal of Physics: Conference Series* have been peer reviewed through processes administered by the proceedings Editors. Reviews were conducted by expert referees to the professional and scientific standards expected of a proceedings journal published by IOP Publishing.

Normally-off GaN Transistors for Power Applications

O Hilt [1], E Bahat-Treidel, F Brunner, A Knauer, R Zhytnytska, P Kotara and J Wuerfl
Ferdinand-Braun-Institut, Leibniz-Institut fuer Hoechstfrequenztechnik
Gustav-Kirchhoff-Strasse 4, 12489 Berlin, Germany

E-mail: oliver.hilt@fbh-berlin.de

Abstract. Normally-off high voltage GaN-HFETs for switching applications are presented. Normally-off operation with threshold voltages of 1 V and more and with 5 V gate swing has been obtained by using p-type GaN as gate. Different GaN-based buffer types using doping and backside potential barriers have been used to obtain blocking strengths up to 1000 V. The increase of the dynamic on-state resistance is analyzed for the different buffer types. The best trade-off between low dispersion and high blocking strength was obtained for a modified carbon-doped GaN-buffer that showed a 2.6x increase of the dynamic on-state resistance for 500 V switching as compared to switching from 20 V off-state drain bias. Device operation up to 200 °C ambient temperature without any threshold voltage shift is demonstrated.

1. Introduction

High voltage GaN-based power switching transistors enable efficient power converters with increased power density. High converter switching frequencies can be realized with lateral GaN-based HFETs due to the low area-specific on-state resistance for a given blocking strength and the low gate charge required for switching [1]. The GaN HFET technology is also considered as cost-efficient since devices can be manufactured on Si substrates using processing steps of a CMOS-Fab environment [2]. Major challenges for market introduction is the conversion of the natural normally-on device characteristic to normally-off and the reduction of the increased dynamic on-state resistance when switching the device from high-voltage off-state bias to the on-state.

Recent attempts to convert AlGaN/GaN high electron mobility transistors (HEMTs) into normally-off devices, using gate recess [3] or fluorine incorporation [4] showed limited applicability for power electronics due to their low threshold voltages $V_{th} < +1$ V, their low gate swing of ~ 2 V and their high on-state gate current. The introduction of a gate insulator suppresses the gate current and may extend the gate swing [2]. The used extrinsic insulator layers are often atomic layer deposited (ALD) oxides like Al_2O_3 or HfO_2 and are deposited at temperatures < 300°C. The observed trap states in the insulator bulk or at the interface to the AlGaN barrier often deteriorate the device switching performance and limit the device reliability [5]. Gate-injection GaN-HFETs using a p-type AlGaN-gate have been developed by Panasonic and give a threshold voltage of +1 V - +2 V [6].

The here presented p-GaN gate GaN-HFETs use a similar concept for normally-off operation as in [6]. While Panasonic highlights the hole-injection regime for on-state operation, this regime is avoided for the presented devices. Hole injection from the gate is paired with additional gate current and an

[1] To whom any correspondence should be addressed.

increased gate charge. Both would contribute to losses during switching transistor operation and should thus get avoided.

This paper focuses on GaN-buffer variations for the lateral GaN devices to obtain an optimum combination of secure normally-off characteristic, high off-state blocking voltage and a low dynamic on-state resistance. Additionally, the high-temperature DC characteristics up to 200 °C base-plate temperature get presented.

2. P-GaN gate normally-off transistors

In the p-GaN gate transistors presented here, the gate consists of in-situ grown Mg-doped GaN with an ohmic contact for biasing [7]. P-GaN gate GaN-HFETs combine the high-mobility two-dimensional electron gas (2DEG) transistor channel known from AlGaN/GaN HEMTs with reliable normally-off operation, as required for applications in power electronics.

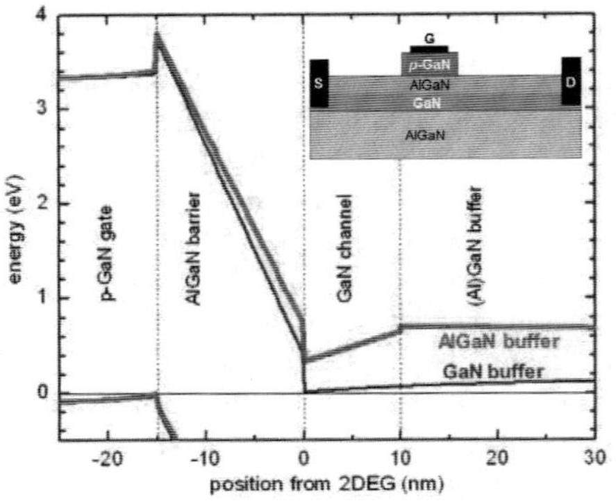

Figure 1. Simulated band diagram at the gate position of a p-GaN gate AlGaN/GaN HFET with GaN buffer (blue) and AlGaN buffer (red). The conduction band energy at the 2DEG position is significantly shifted above the Fermi level for the AlGaN buffer structure. The device structure is sketched in the inset.

The gate acceptors deplete the 2DEG of the transistor channel, when unbiased and a positive gate bias is needed to open the transistor. The simulated band diagram at the gate position (figure 1) confirms that the 2DEG channel at the heterojunction between GaN channel and AlGaN barrier does not cross the Fermi level (blue conduction band line in figure 1). Replacing the GaN buffer by AlGaN and using uid GaN for the transistor channel generates a virtual p-type doping at the AlGaN back-barrier of the channel and further increases the 2DEG depletion beneath the gate (red conduction band line in figure 1). Fabricated p-GaN gate transistors with $Al_{0.05}Ga_{0.95}N$ buffer and with iron-doped buffer (GaN:Fe) demonstrate the normally-off characteristics, see figure 2. Threshold voltages of +1 V and above have been realized for this normally-off approach.

In on-state, the gate bias can be driven to 5 V and more. The nature of the pin-type gate diode restricts the on-state gate current to 10 μA/mm for $V_{GS} = 5$ V. Driving the gate to $V_{GS} > 7$ V will lead to hole injection from the gate to the transistor channel as described in [6] for the p-AlGaN gate transistors.

In off-state, at $V_{GS} = 0$ V, the drain leakage current is limited to 1 μA/mm for the AlGaN-buffer device and to 30 μA/mm for the GaN:Fe-buffer device. As suggested by figure 1, a higher threshold voltage has been obtained for the AlGaN buffer devices with the incorporated back barrier.

250 V / 85 mΩ normally-off switching transistors in p-GaN gate technology have been fabricated by using the AlGaN buffer approach. The required device gate width of 113 mm has been divided into 30 sub-cells that are connected via bumps to the sub-mount (right side of figure 3).

Journal of Physics: Conference Series **494** (2014) 012001 doi:10.1088/1742-6596/494/1/012001

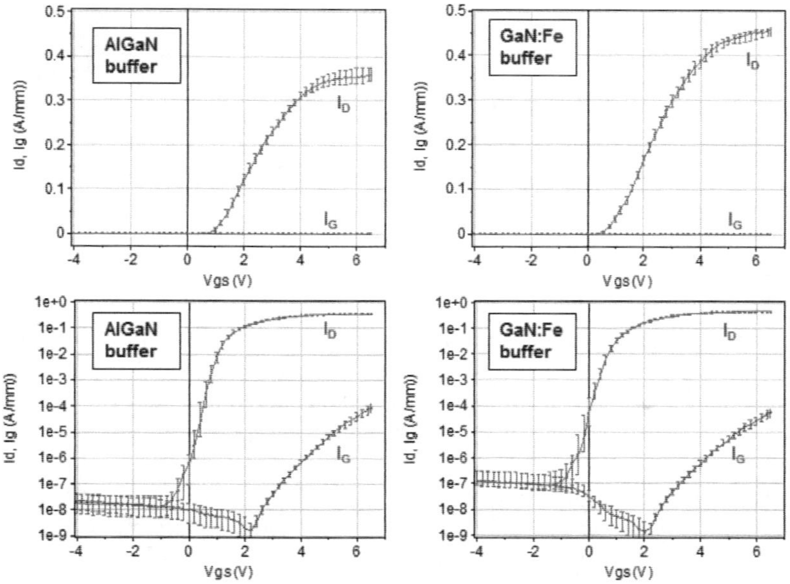

Figure 2. Transfer characteristics and gate current (linear representation on top, logarithmic representation on bottom) of p-GaN gate HFETs based on an AlGaN buffer (left) and on an iron-doped GaN buffer (right). The median currents with 25%/75%-quantiles (error bars) of 30 devices spread over a 3" wafer are shown. Device gate width is 2.1 mm, gate drain separation is 15 µm and V_{DS} = 10 V.

The device output characteristics demonstrate off-state condition for V_{GS} = 1 V and on-state condition for V_{GS} = 5 V (left side of figure 3). The current can be driven to more than 20 A. The 250 V blocking strength is derived from the off-state drain leakage < 1 mA/mm. The area-specific on-state resistance is $R_{ON}A = 1\ \mathrm{m\Omega cm^2}$ with considering the active device area as A only. This value is approx. a factor 5 lower than for Si-based MOSFETs of the same voltage class [7].

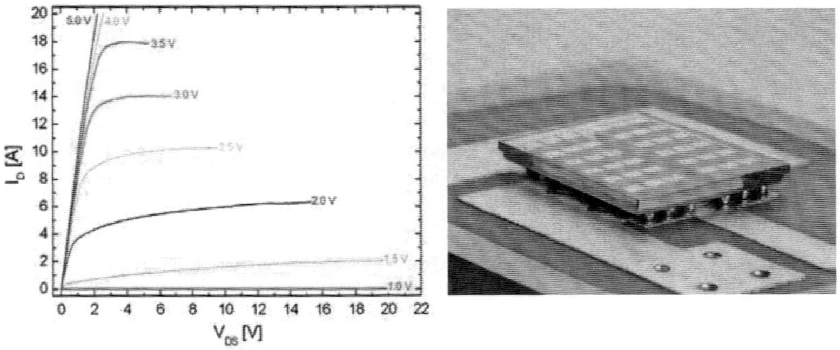

Figure 3. Output characteristics (left) and photograph (right) of a 250 V / 85 mΩ normally-off GaN HFET in p-GaN gate technology. The device has 113 mm gate width and is assembled in flip-chip technology.

3. Buffer concepts for high voltage operation

The breakdown voltage of GaN HFETs should ideally increase with the gate-drain separation d_{GD} since the electric field in off-state distributes between gate and drain. This breakdown voltage scaling is lost when electrons from the transistor channel manage to bypass the gate control region via deep layers in the buffer. The resulting source-drain leakage current prevents higher blocking voltages, independent of d_{GD}. The major cause for this electron punch-through effect is the limited barrier height of the GaN buffer beneath the transistor channel with the two-dimensional electron gas [8]. The Fermi level in uid-doped MOCVD-grown GaN-layers is usually shifted towards the conduction band and the back-side barrier towards the buffer is in the order of 100 meV, see figure 1.

Punch-through effects and the associated source-drain leakage can get suppressed by a better electron confinement to the channel. An AlGaN buffer beneath the GaN channel creates an efficient barrier for the channel electrons (figure 1). The buffer conduction band is shifted upwards due to the polarization differences and the band energy differences between the GaN channel and the AlGaN buffer [8]. A breakdown voltage scaling of 40 V/μm gate-drain separation was measured for $Al_{0.05}Ga_{0.95}N$ back-barrier based p-GaN gate devices and 870 V breakdown voltage has been obtained [7]. GaN-buffer compensation doping with carbon (GaN:C) or iron (GaN:Fe) also shifts the buffer Fermi level towards the valence band and prevents electrons in the channel to enter the buffer. More than 1000 V breakdown strength with 100 V/μm scaling has been obtained for a GaN:C-based p-GaN gate transistor [10]. GaN:Fe buffer yield 50 V/μm blocking strength scaling [9].

4. Dynamic characteristics and high voltage switching

An increase of the dynamic on-state resistance (dyn. R_{ON}) immediately after switching from high-bias off-state condition is often observed in GaN-HFETs. Trap states in the (Al)GaN-based semiconductor stack or on the semiconductor surface are considered as root cause for this effect [11]. Electron traps inside the buffer may get filled under high bias off-state condition and deplete the transistor channel after switching the transistor to on-state.

Buffer material with very low trap densities is thus considered as straightforward for low-dispersion switching properties of the transistor. Non-compensated buffer designs, however, may be in conflict to the buffer optimization approaches as discussed above for optimum device blocking strength. Pure GaN buffer suffer from punch-through and the device blocking strength of the (non-doped) AlGaN buffer devices is only mediocre.

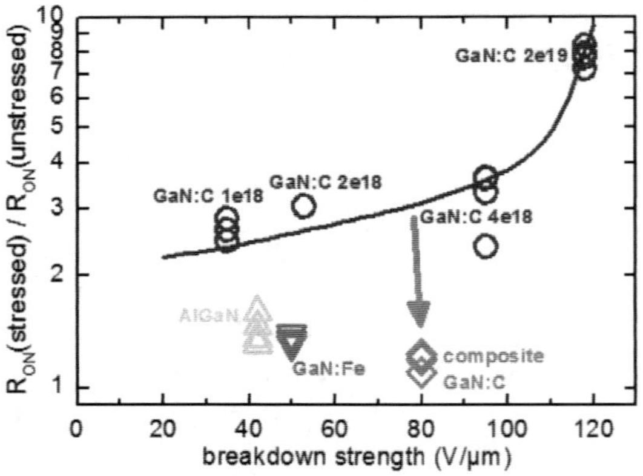

Figure 4. Increase in dynamic on-state resistance for 0.2 μs long pulses into on-state from 65 V off-state bias (with respect to switching from 0 V drain bias) in relation to the device breakdown strength scaling. Test transistors with 0.25 mm gate width and different buffer compositions are compared.

In fact, the increase in dyn. R_{ON} after 65 V off-state bias is for the AlGaN-buffer based devices (42 V/μm blocking strength) much lower than for the 2e19 cm^{-3}-doped GaN:C buffer devices with 120 V/μm blocking strength, see figure 4. Lower carbon doping reduces both, the increase in dynamic R_{ON} but also the blocking strength. Using iron as buffer dopant results in a better trade-off between dyn. R_{ON} increase and blocking strength as compared to the AlGaN-buffer and GaN:C-buffer devices.

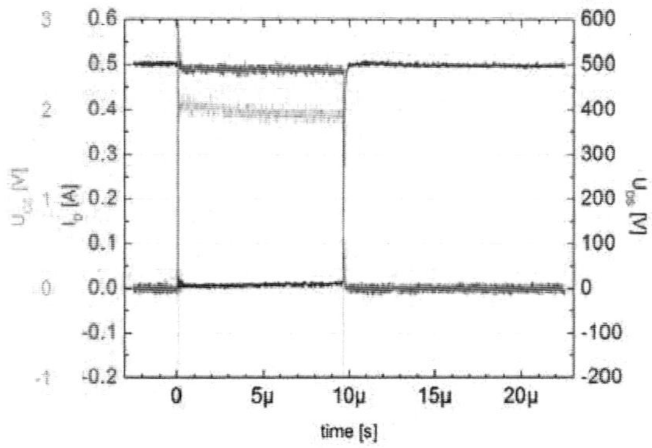

Figure 5. 500 V / 0.5 A switching transient for a GaN HFET with modified GaN:C buffer structure and 2.1 mm gate width. The dynamic R_{ON}, as extracted from the on-state voltage drop is 16 Ω, corresponding to a 2.6x increase of the R_{ON}.

For devices with 50 V/µm blocking strength, the increase in dyn. R_{ON} is factor 3 for the 2e18 cm^{-3} -doped GaN:C-buffer but only factor 1.4 for the GaN:Fe-buffer.

The complete semiconductor stack charge balance and the resulting potentials inside the device have to be taken into account for a deeper understanding of the dispersion phenomena [12]. Recent switching experiments with a modified composite GaN:C-buffer device that additionally had an AlGaN back-barrier and a GaN cap incorporated showed a successful combination of high blocking strength and low dispersion. The device with 80 V/µm blocking strength showed only a 1.2-times increase in dyn. R_{ON} for pulsing from 65 V (figure 4). Drain voltage and drain current transients during 500 V / 0.5 A switching with a packaged device were taken (figure 5). The measured dynamic R_{ON} was 16 Ω for the device with 2.1 mm gate width. This corresponds to an 2.6x increase in dyn. R_{ON} after switching from 500 V.

5. Temperature dependence

GaN-based switches can in principle operate at elevated temperatures of 300°C or more due to their high bandgap energy of 3.4 eV. The temperature dependence of the drain current and of the threshold voltage is important for the thermal stability of the device as thermal runaway can be avoided by a negative temperature coefficient of the drain current. Figure 6 shows the transfer and on-state output characteristics at ambient temperatures between 25 °C and 200 °C for small test transistors with low self-heating in the on-state.

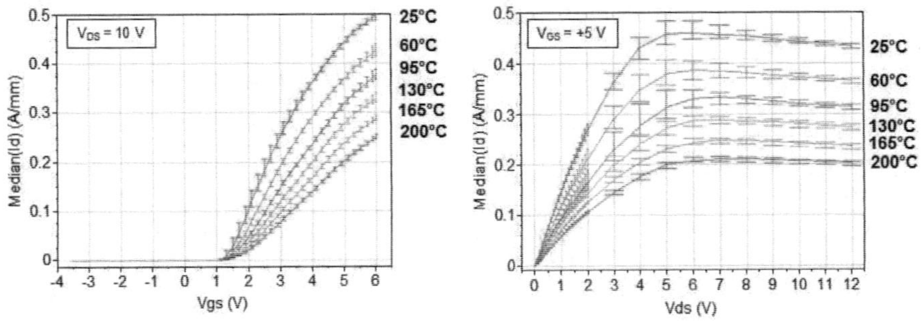

Figure 6. Temperature dependence of the transfer characteristic (left) and the on-state IV curve (right) of 30 p-GaN gate HFETs on one 3" wafer. The median currents with 25%/75%-quantiles (error bars) are shown. Gate width is 0.25 mm and gate drain separation is 6 µm.

The threshold voltage is within a variation of +/- 80 mV essentially independent on temperature. The on-state resistance increases with temperature ($\alpha = 8.2 \times 10^{-3}$ K^{-1}) and the maximum drain current in on-state shows a negative temperature coefficient $\alpha = -2.9 \times 10^{-3}$ K^{-1}. Stability of the threshold voltage is due to the essentially temperature-independent electron concentration in the 2DEG, the temperature dependent saturation current (and R_{ON}) is related to the phonon-scattering of the electrons, increasing with temperature [13].

There is no device operation point with a positive α for the drain current due to the temperature-independent V_{th}. Current focusing and hot spots on limited sub cells of the transistor get thus also avoided for operation conditions with not fully opened transistor.

6. Conclusion

250 V / 85 mΩ normally-off p-GaN gate transistors with 1 V threshold voltage and 5 V gate swing have been demonstrated. The threshold voltage is stable up to 200 °C. The device buffer structure has to be designed carefully for high-voltage operation and efficient switching. 500 V / 0.5 A switching with an only 2.6x increased dynamic on-state resistance has been demonstrated.

7. References

[1] Saito W, Nitta T, Kakiuchi Y, Saito, Tsuda K, Omura I and Yamaguchi M 2008 *IEEE Electr. Device Lett.* **29**, 8

[2] De Jaeger B, Van Hove M, Wellekens D, Kang X, Liang H, Mannaert G, Geens K and Decoutere S 2012 *24th Int. Symp. on Power Semiconductor Devices and ICs (ISPSD), Bruges, Belgium, June 3-7* pp 49-52

[3] Saito J, Takada Y, Kuraguchi M, Tsuda K and Omura I 2006 *IEEE Trans. on Electron Devices* **53** 356

[4] Cai Y, Zhou Y, Lau K and Chen K 2006 *IEEE Trans. on Electron Devices* **53** 2207

[5] Sun X, Saadat O, Chang-Liao K, Palacios T, Cui S and Ma T 2013 *Appl. Phys. Lett.* **102** 103504

[6] Uemoto Y , Hikita M, Ueno H, Matsuo H, Ishida H, Yanagihara M, Ueda T, Tanaka T and Ueda D 2007 *IEEE Trans. on Electron Devices* **54** 3393

[7] Hilt O, Knauer A, Brunner F, Bahat-Treidel E and Würfl J 2010 *Proc. 22nd Int. Symp. on Power Semiconductor Devices & ICs (ISPSD), Hiroshima, Japan, June 6-10* pp 347-350

[8] Bahat-Treidel E, Hilt O, Brunner F, Würfl J, and Tränkle G 2008 *IEEE Trans. on Electron Devices* **55** 3354

[9] Hilt O, Kotara P, Brunner F, Knauer A, Zhytnytska R, and Würfl J 2013 *IEEE Trans. on Electron Devices* **60** 3084

[10] Hilt O, Bahat-Treidel E, Cho E, Singwald S and Würfl J 2012 *24th Int. Symp. on Power Semiconductor Devices and ICs (ISPSD), Bruges, Belgium, June 3-7*, pp 345-348

[11] Jin D and del Alamo J 2012 *24th Int. Symp. on Power Semiconductor Devices and ICs (ISPSD), Bruges, Belgium, June 3-7* pp 333-336

[12] Uren M, Möreke J and Kuball M 2012 *IEEE Trans. on Electron Devices* **59** 3327

[13] Pérez-Tomás A, Placidi M, Baron N, Chenot S, Cordier Y, Moreno J, Constant A, Godignon P and Millán J 2009 *J. Appl. Phys.* **106** 074519

Acknowledgements

Support from the German BMBF (LES project PowerGaNPlus, contract no. 13N10908) and from EU FP7 project HipoSwitch (Grant no. 287602) is gratefully acknowledged.

Scaling influence on the thermal behavior of toward-THz SiGe:C HBTs

V d'Alessandro[1], G Sasso[1], N Rinaldi[1] and K Aufinger[2]
[1] Department of Electrical Engineering and Information Technology, University Federico II, via Claudio 21, 80125 Naples, Italy
[2] Infineon Technologies AG, Am Campeon 1-12, 85579 Neubiberg, Germany

E-mail: vindales@unina.it

Abstract. An extensive on-wafer experimental campaign is performed to extract the thermal resistance of state-of-the-art toward-THz silicon germanium bipolar transistors designed and developed within the European DOTFIVE project. The dependence of this critical parameter on scaling, as well as on the emitter layout, is carefully evaluated, and the resulting junction temperatures are determined.

1. Introduction
Silicon germanium (SiGe) bipolar technology is increasingly adopted in a large variety of mm-wave and near-THz applications, like high-bandwidth communications, optical transmission, medical equipments, and automotive radars. An important contribution to this trend in the European scenario has been provided by the recently-ended DOTFIVE project [1] aimed at demonstrating heterojunction bipolar transistors (HBTs) with maximum oscillation frequency f_{max}=0.5 THz [2]-[4], which the new-born DOTSEVEN [5] is expected to boost to 0.7 THz. The need for improving the performance of SiGe transistors has motivated a constant technological effort to allow aggressive lateral and vertical scaling. However, this is leading to unsustainable thermal issues induced by (i) the increase in current (and power) density, as clearly stated in a comprehensive review paper [6], and (ii) the reduction of the spacing between the intrinsic region and shallow/deep trenches filled with low thermal conductivity materials [7], for which the thermal resistances of single-finger transistors have been pushed into the thousands of K/W [8], [9].

This paper contributes to deepen the understanding of the scaling impact on the self-heating (SH) in SiGe:C HBTs by means of simple DC measurements carried out on state-of-the-art devices realized in the DOTFIVE framework.

2. Devices Under Test and Experimental Setup
Measurements were performed on about 100 single-emitter SiGe:C NPN transistors manufactured by Infineon Technologies. The devices are divided into three sets corresponding to different technology stages of the project development, which are hereinafter denoted as #1, #2, and #3. For each set, HBTs with several combinations of emitter width/length were available.

The sets are characterized by different lateral/vertical scaling levels. In particular,

- Set #2 is slightly laterally/vertically scaled compared to #1; the collector current of HBTs belonging to #2 at maximum cut-off frequency f_T is ~30% higher than that of the #1 counterparts with approximately the same emitter area.

Journal of Physics: Conference Series **494** (2014) 012002 doi:10.1088/1742-6596/494/1/012002

- Set #3 transistors underwent an aggressive lateral scaling compared to #2 devices, while being vertically similar to them.

The key figures of the sets, i.e., the open-emitter breakdown voltage BV_{CBO}, the peak frequencies f_T, f_{max} at V_{CB}=0 V, and the current density J_C at peak f_T (extracted for transistors with small W_E), are reported in table 1.

Table 1. Key figures of the analyzed HBTs.

Set	BV_{CBO} [V]	peak f_T [GHz]	J_C @ peak f_T [mA/μm^2]	peak f_{max} [GHz]
#1	6.5-6.8	190	6.5-7.0	250
#2	5.2-5.9	230	9.0-9.5	310
#3	5.1-5.5	235	9.5-10	330

All the devices under test (DUTs) have single base and collector contacts (BEC configuration), as illustrated by the schematic cross-section reported in figure 1, which also evidences some key geometric parameters, like the effective emitter width W_E, the width $W_{P\text{-}poly}$ of the P^+ polysilicon window (i.e., the distance between the P^+ polysilicon edges along *x*), and the widths of the silicon boxes surrounded by deep (W_{DTbox}) and shallow (W_{STbox}) trenches. The mask (drawn) emitter width $W_{E,drawn}$ is equal to $W_{P\text{-}poly}$ for sets #1 and #2, and to $W_{P\text{-}poly}$-20 nm for set #3. Similar definitions hold for the lengths along *y*. The thicknesses (along *z*) of the shallow and deep trenches are equal to 0.3 μm and ~4.5 μm, regardless of the set.

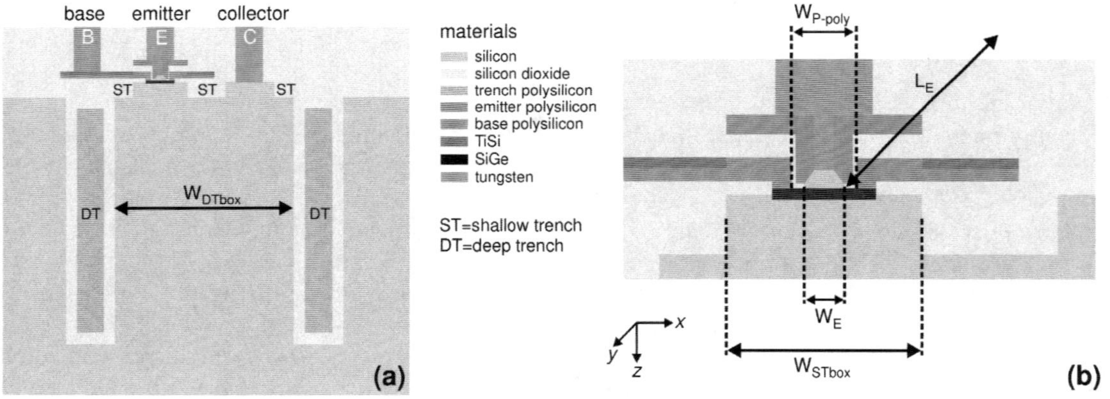

Figure 1. (a) Sketch of the cross-section of the typical HBT under test and (b) magnification of the intrinsic transistor region.

The on-wafer experimental campaign was carried out by means of a PM5 Karl Suss probe station. The electrical signals were handled by an HP4142B parameter analyzer. DC common-emitter (CE) and common-base (CB) measurements were performed by using PH100 probeheads equipped with tungsten needles. Countermeasures were taken to prevent the (sometimes destructive) oscillations that may arise under CB conditions; in particular, ferrite beads were applied to the cables in the close proximity of the DUTs. The baseplate (thermochuck) temperature T_B was set to assigned values thanks to an ATT heating/cooling system.

3. Thermal Resistance Extraction Approach

The extraction method is based on DC measurements only, and can be considered as an improved CB variant of the classic approach proposed by Dawson *et al.* [10]. The procedure is articulated in two steps, namely, (1) the calibration of the temperature coefficient of the base-emitter voltage, which is employed as a *thermometer*, and (2) the evaluation of the slope of a CB I_E-constant V_{BE}–V_{CB} characteristic.

3.1. Transistor model

If the collector-base voltage V_{CB} is limited to values sufficiently low to neglect impact-ionization (II), the collector current can be expressed as

$$I_C = A_E \cdot J_S \cdot \exp\left(\frac{V_{BEj} + \phi \cdot \Delta T_j}{\eta \cdot V_{T0}}\right) = A_E \cdot J_S \cdot \exp\left(\frac{V_{BE} - R_{EB} \cdot I_C + \phi \cdot \Delta T_j}{\eta \cdot V_{T0}}\right) \tag{1}$$

where

- $A_E = W_E \times L_E$ [μm^2] is the effective emitter area;
- J_S [$A/\mu m^2$] is the reverse saturation current density;
- η is the ideality factor;
- V_{T0} (=25.86 mV) is the thermal voltage at ambient temperature (T_0=300 K);
- V_{BEj} and V_{BE} [V] are the internal (junction) and the externally accessible base-emitter voltages, respectively;
- $\Delta T_j = T_j - T_0$ [K] is the (average) temperature rise above ambient over the base-emitter junction;
- ϕ [V/K] is the temperature coefficient of V_{BEj} (in absolute value), i.e.,

$$\phi = -\frac{\partial V_{BEj}}{\partial T_j} \tag{2}$$

- R_{EB} [Ω] is an "aggregate" resistance dependent on the parasitic base (R_B) and emitter (R_E) resistances, as well as on the CE current gain β_F, given by

$$R_{EB} = \frac{R_B + R_E}{\beta_F} + R_E \approx \frac{R_B}{\beta_F} + R_E \tag{3}$$

3.2. Thermometer calibration

The technological coefficient ϕ was evaluated as follows. The DUTs were mounted in a CB configuration and the V_{BE}–T_B characteristics were measured at assigned I_E and V_{CB}, the values of which were chosen sufficiently low to annihilate SH, II, and resistive effects, so that $T_j \approx T_B$ and $V_{BEj} \approx V_{BE}$. As a result, the V_{BE}–T_B curve coincides with the V_{BEj}–T_j one, and ϕ can be obtained from its slope. By repeating the extraction procedure for various I_E values, it was found that ϕ logarithmically depends on I_E (almost equal to the II-free I_C), as already found for GaAs [11] and Si [12] bipolar transistors, as well as for SiGe:C devices fabricated by STMicroelectronics [8]. This dependence can be described by the following relationship:

$$\phi(I_E) = \phi_0 - \frac{k}{q} \cdot \ln\left(\frac{I_E}{A_E \cdot J_S}\right) \tag{4}$$

where k is the Boltzmann's constant and q is the elementary positive charge. Parameter ϕ_0 – an expression of which as a function of device parameters can be derived from the analysis developed in [12] – was determined by fitting (4) with the experimental ϕ vs. I_E data. It was found that ϕ_0 is set-dependent, but almost layout-insensitive for a given set; the extracted values are ϕ_0=3.53, 3.47, and 3.37 mV/K for sets #1, #2, and #3, respectively. The accuracy of the ϕ_0 estimation is witnessed by the favorable agreement at low-medium current levels between the experimental I_C–V_{BE} curves measured at various T_B for a low V_{CE} and the first-order model given by

$$I_C = A_E \cdot J_S \cdot \exp\left[\frac{V_{BE} + \phi_0 \cdot (T_B - T_0)}{\eta V_{T0} + (k/q) \cdot (T_B - T_0)}\right] \tag{5}$$

which can be obtained by substituting the logarithmic ϕ relationship (4) (with the II-free I_C in place of I_E) into the transistor model (1) and disregarding SH and resistive effects; this is shown in figure 2 for devices belonging to all sets.

From this analysis, it was also found that the scaling leads to an increase in J_S (J_S=1.7×10^{-16} A/µm^2 for set #1, J_S=2.8×10^{-16} A/µm^2 for #2, and J_S=3.1×10^{-16} A/µm^2 for #3) and a reduction in η (η=1.08-1.1 for #1, η=1.06 for #2, and η=1.035-1.055 for #3).

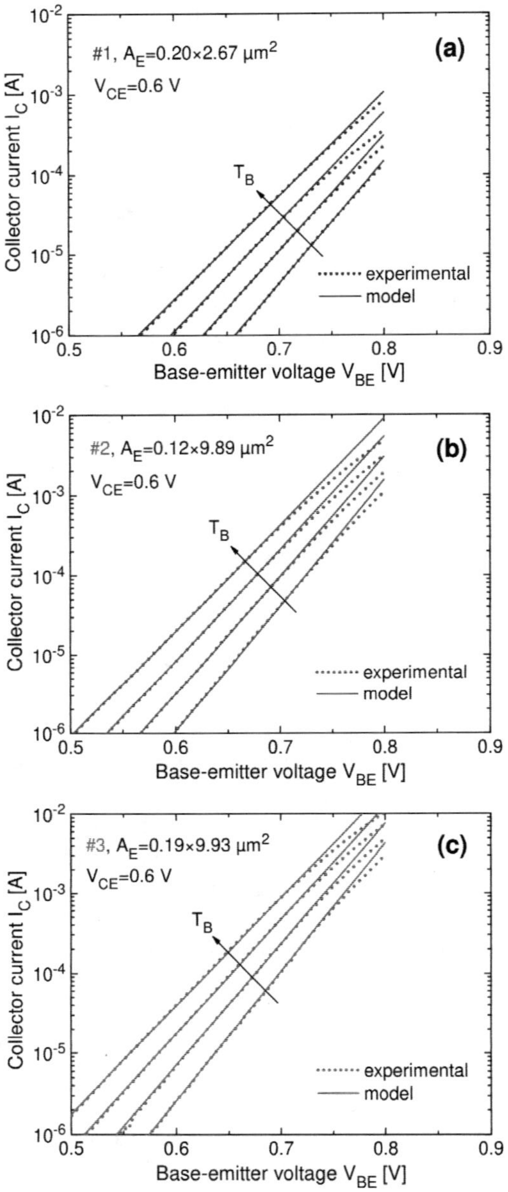

Figure 2. Collector current I_C vs. base-emitter voltage V_{BE} at V_{CE}=0.6 V and T_B=300, 320, 340, 360 K for three HBTs belonging to sets (a) #1, (b) #2, and (c) #3; experimental data (dotted lines) are compared with the calibrated model (5) (solid).

3.3. Thermal resistance evaluation

From (1), if the II-free I_C is replaced by I_E, V_{BE} can be expressed as

$$V_{BE} = R_{EB} \cdot I_E - \phi(I_E) \cdot \Delta T_j + \eta \cdot V_{T0} \cdot \ln\left(\frac{I_E}{A_E \cdot J_S}\right) \tag{6}$$

where ΔT_j is related to the dissipated power P_D by the thermal equivalent of the Ohm's law

$$\Delta T_j = R_{TH} \cdot P_D = R_{TH} \cdot \left(V_{BE} \cdot I_E + V_{CB} \cdot I_C\right) \approx R_{TH} \cdot \left(V_{BE} + V_{CB}\right) \cdot I_E \tag{7}$$

R_{TH} being the thermal resistance of the device. By substituting (7) into (6), it is found that

$$V_{BE} = \frac{1}{1+\phi(I_E)R_{TH}I_E}\cdot\left[R_{EB}I_E - \phi(I_E)R_{TH}I_E V_{CB} + \eta\cdot V_{T0}\cdot\ln\left(\frac{I_E}{A_E\cdot J_S}\right)\right] \tag{8}$$

which allows obtaining the slope γ of an I_E-constant V_{BE}–V_{CB} characteristic as

$$\gamma = \frac{dV_{BE}}{dV_{CB}} = \frac{I_E}{1+\phi(I_E)R_{TH}I_E}\cdot\left[\frac{dR_{EB}}{dV_{CB}} - \phi(I_E)R_{TH}\right] \tag{9}$$

In order to evaluate dR_{EB}/dV_{CB}, it must be considered that R_{EB} – given by (3) – depends on the junction temperature T_j through (i)

$$R_B = R_{B0}\cdot\left(\frac{T_j}{T_0}\right)^m \qquad R_E = R_{E0}\cdot\left(\frac{T_j}{T_0}\right)^m \tag{10}$$

where R_{B0} and R_{E0} are the resistance values at $T=T_0$ and m (>0) is a power factor assumed identical for both, and (ii)

$$\beta_F = \beta_{F0}\cdot\exp\left[\frac{\Delta E_G}{k}\cdot\left(\frac{1}{T_j}-\frac{1}{T_0}\right)\right] \tag{11}$$

where β_{F0} is the CE current gain at $T=T_0$, and ΔE_G is the difference between the bandgaps of the Si emitter and SiGe base.[1] The temperature T_j is in turn dependent upon V_{BE} and V_{CB} via (7), and therefore R_{EB} can be considered as a function of V_{BE} and V_{CB}; by differentiating,

$$dR_{EB} = \left.\frac{\partial R_{EB}}{\partial V_{BE}}\right|_{V_{CB}} dV_{BE} + \left.\frac{\partial R_{EB}}{\partial V_{CB}}\right|_{V_{BE}} dV_{CB} \tag{12}$$

from which

$$\frac{dR_{EB}}{dV_{CB}} = \left.\frac{\partial R_{EB}}{\partial V_{BE}}\right|_{V_{CB}} \frac{dV_{BE}}{dV_{CB}} + \left.\frac{\partial R_{EB}}{\partial V_{CB}}\right|_{V_{BE}} \tag{13}$$

By replacing (13) into (9), it is obtained that

$$\gamma = \frac{dV_{BE}}{dV_{CB}} = \frac{I_E}{1+\phi(I_E)R_{TH}I_E}\cdot\left[\left.\frac{\partial R_{EB}}{\partial V_{BE}}\right|_{V_{CB}} \frac{dV_{BE}}{dV_{CB}} + \left.\frac{\partial R_{EB}}{\partial V_{CB}}\right|_{V_{BE}} - \phi(I_E)R_{TH}\right] \tag{14}$$

and hence

$$\gamma = \frac{dV_{BE}}{dV_{CB}} = -\frac{\phi(I_E)R_{TH}I_E - I_E\cdot\left.\frac{\partial R_{EB}}{\partial V_{CB}}\right|_{V_{BE}}}{1+\phi(I_E)R_{TH}I_E - I_E\cdot\left.\frac{\partial R_{EB}}{\partial V_{BE}}\right|_{V_{CB}}} \tag{15}$$

After simple algebra, it can be found that

$$\left.\frac{\partial R_{EB}}{\partial V_{CB}}\right|_{V_{BE}} = \left.\frac{\partial R_{EB}}{\partial V_{BE}}\right|_{V_{CB}} = R_{TH}I_E\frac{\partial R_{EB}}{\partial T_j} \tag{16}$$

and (15) becomes

$$\gamma = \frac{dV_{BE}}{dV_{CB}} = -\frac{R_{TH}I_E\cdot\left[\phi(I_E) - I_E\cdot\frac{\partial R_{EB}}{\partial T_j}\right]}{1+R_{TH}I_E\cdot\left[\phi(I_E) - I_E\cdot\frac{\partial R_{EB}}{\partial T_j}\right]} \tag{17}$$

[1] Equation (11) can be easily derived by reasonably assuming that the main temperature dependence of β_F comes from $(n_{iB}/n_{iE})^2$, n_{iB} and n_{iE} being the intrinsic carrier concentrations in the base and emitter, respectively.

The derivative $\partial R_{EB}/\partial T_j$ can be determined by making use of (3), (10), and (11), and is given by

$$\frac{\partial R_{EB}}{\partial T_j}=\left(\frac{T_j}{T_0}\right)^m\cdot\left[\frac{R_{B0}}{\beta_F\cdot T_j^2}\cdot\left(m\cdot T_j+\frac{\Delta E_G}{k}\right)+\frac{m\cdot R_{E0}}{T_j}\right] \tag{18}$$

It must be remarked that (18) reduces to the formula presented in [12] for Si bipolar transistors by neglecting the temperature dependence of the parasitic resistances R_E and R_B (m=0), and replacing ΔE_G with $-\Delta E_{G(BGN)}$, where $\Delta E_{G(BGN)}$ represents the difference between the bandgaps of the base and the emitter, the latter being impacted by bandgap narrowing due to the high doping level. Since $\partial R_{EB}/\partial T_j \ll 1$ for typical parameter values, $I_E\cdot\partial R_{EB}/\partial T_j$ can be neglected with respect to ϕ without loss of accuracy in (17), which reduces to [12]

$$\gamma=\frac{-R_{TH}\cdot I_E\cdot\phi(I_E)}{1+R_{TH}\cdot I_E\cdot\phi(I_E)}\approx -R_{TH}\cdot I_E\cdot\phi(I_E) \tag{19}$$

whence

$$R_{TH}=-\frac{\gamma}{I_E\cdot\phi(I_E)} \tag{20}$$

The thermal resistance can be therefore determined by (i) measuring a V_{BE}–V_{CB} characteristic at I_E high enough to give rise to perceptible SH, the V_{CB} range being limited to values at which II can be reasonably disregarded, (ii) extracting the slope γ of this curve (see figure 3 for an example), and (iii) invoking (20), where the value of coefficient $\phi(I_E)$ is calculated from (4) for each HBT.

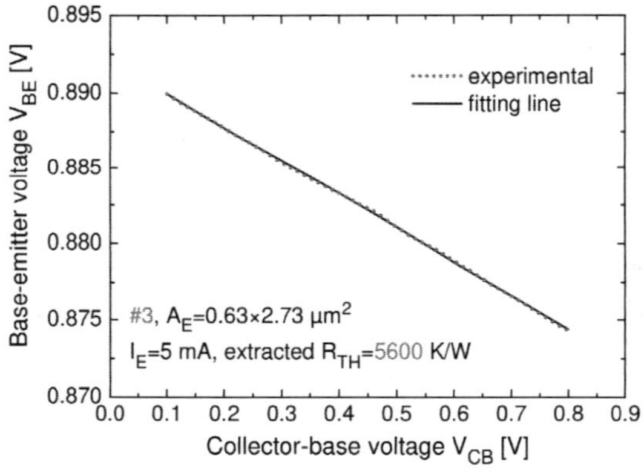

Figure 3. Extraction of the slope γ of an I_E-constant V_{BE}–V_{CB} characteristic for a device belonging to set #3.

4. Results and Discussion

Figure 4 illustrates R_{TH} as a function of effective emitter length L_E at assigned widths W_E for several HBTs belonging to all sets. It is shown that R_{TH} significantly increases by reducing L_E and can grow beyond 10^4 K/W for small areas. This confirms the findings in [8], where some CBEBC devices fabricated by STMicroelectronics were characterized by using the DC extraction technique employed in this work, and in [9], where the thermal impedances were determined by resorting to small-signal S-parameter measurements. In particular, the smallest DUTs belonging to set #1 (A_E=0.2×0.57 µm^2), #2 (A_E=0.14×0.39 µm^2), and #3 (A_E=0.11×0.63 µm^2) suffer from R_{TH}=14300 K/W, R_{TH}=21000 K/W, and R_{TH}=22000 K/W, respectively.

In figure 5, a comparison between sets is carried out for the same values of mask emitter length by varying the emitter width. The sharp R_{TH} growth for shrunk devices is apparent. It is inferred that the resistances of set #2 are slightly higher than those corresponding to #1, which can be attributed to (i) the reduced spacing between active area and shallow trench that contributes to confine the heat flow within the intrinsic transistor region, and, to a lower extent, (ii) the vertical shrinking of the heat source (i.e., the power-dissipating collector-base space charge region) induced by the smaller epi thickness and higher collector doping, which increases the volumetric power density [13]. A more marked increment in R_{TH} is observed from set #2 to #3; this is ascribable to (i) the aggressive lateral scaling leading to a significant decrease in both W_{STbox} and W_{DTbox}, which effectively counteracts the lateral heat propagation, and (ii) the reduced cooling action of the narrowed emitter contact.

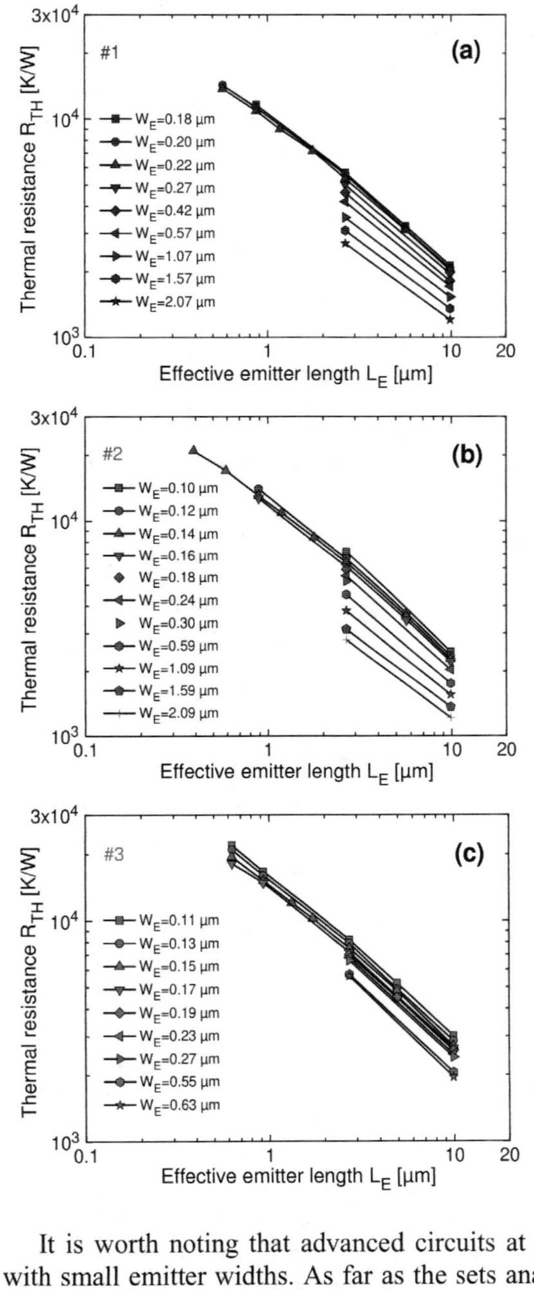

Figure 4. Thermal resistance R_{TH} as a function of effective emitter length L_E for various effective emitter widths W_E, as experimentally determined for sets (a) #1, (b) #2, and (c) #3.

It is worth noting that advanced circuits at a given technology generation are typically designed with small emitter widths. As far as the sets analyzed in this work are concerned, the best candidates are W_E=0.2 μm ($W_{E,drawn}$=0.33 μm) for set #1, W_E=0.14 μm ($W_{E,drawn}$=0.25 μm) for #2, and W_E=0.13 μm ($W_{E,drawn}$=0.2 μm) for #3. Thus, from a practical viewpoint, it is more meaningful to compare the thermal behavior of DUTs sharing such W_E values. Figure 6a shows that – with respect to set #1 – R_{TH} increases by 23.5% (#2) and 52.5% (#3) for short HBTs ($L_{E,drawn}$=0.7 μm) and by 11.2% (#2) and 39% (#3) for long ones ($L_{E,drawn}$=10 μm). Figure 6b reports the corresponding junction temperature rises ΔT_j, as determined by multiplying the thermal resistances by the power dissipated by the devices at peak f_T for V_{CE}=1 V; current density values J_C=6.8, 9, and 9.5 mA/μm^2 were reasonably chosen as representative of sets #1, #2, and #3, respectively. It is found that the scaling-induced ΔT_j increase is almost L_E-independent, and amounts to about 15% from set #1 to #2, and to 25% from #2 to #3.

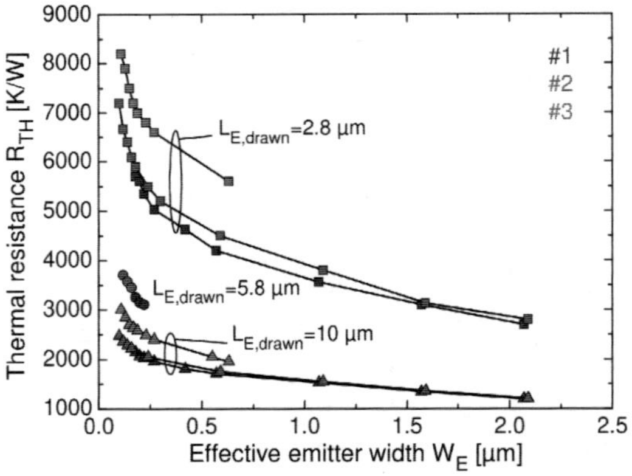

Figure 5. Thermal resistance R_{TH} as a function of effective emitter width W_E; data evaluated for the three sets under test are compared for the same values of mask (drawn) emitter length $L_{E,drawn}$.

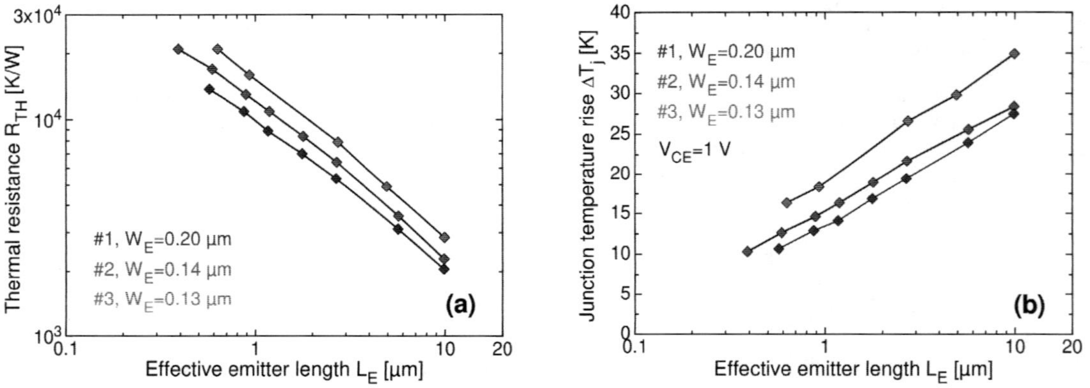

Figure 6. (a) Thermal resistance R_{TH} and (b) junction temperature rise above ambient ΔT_j as a function of effective emitter length L_E; the data correspond to the W_E values adopted for advanced circuit design.

5. Conclusions

In this paper, the influence of scaling and emitter width/length on the thermal behavior of state-of-the-art SiGe:C HBTs has been investigated through a wide experimental analysis carried out on three technology stages developed within the framework of the European DOTFIVE project. A simple extraction procedure has been exploited, which is based on DC measurements and makes use of a carefully calibrated thermometer relating the base-emitter voltage to the junction temperature at each current level. The main results can be summarized as follows: (i) small-area devices suffer from a thermal resistance R_{TH} exceeding 10^4 K/W; (ii) R_{TH} can increase by up to 50% for transistors to be employed in advanced circuits when aggressive lateral scaling is performed; (iii) the concurrent scaling-triggered growth in thermal resistance and current density leads to a significant increase in junction temperature. This study allows foreseeing that the technology improvements will paradoxically pose a serious threat to the reliability and performance of the transistors, if adequate countermeasures are not taken.

Acknowledgment
The authors wish to acknowledge the DOTSEVEN project (316755) supported by the European Commission through the Seventh Framework Programme (FP7) for Research and Technology Development.

References

[1] http://www.dotfive.eu/
[2] Chevalier P, Pourchon F, Lacave T, Avenier G, Campidelli Y, Depoyan L, Troillard G, Buczko M, Gloria D, Céli D, Gaquière C and Chantre A 2009 *Proc. IEEE BCTM (Capri, Italy)* 1
[3] Al-Sa'di M, d'Alessandro V, Frégonèse S, Hong S-M, Jungemann C, Maneux C, Marano I, Pakfar A, Rinaldi N, Sasso G, Schröter M, Sibaja-Hernandez A, Tavernier C and Wedel G 2010 *Proc. IEEE EuMIC (Paris, France)* 29
[4] Chevalier P, Meister T F, Heinemann B, Van Huylenbroeck S, Liebl W, Fox A, Sibaja-Hernandez A and Chantre A 2011 *Proc. IEEE BCTM (Atlanta, Georgia, USA)* 57
[5] http://www.dotseven.eu/
[6] Cressler J D 2013 *Proc. IEEE SiRF (Austin, Texas, USA)* 81
[7] Rieh J-S, Greenberg D, Liu Q, Joseph A J, Freeman G and Ahlgren D C 2005 *IEEE Trans. Electron Devices* **52** 2744
[8] d'Alessandro V, Marano I, Russo S, Céli D, Chantre A, Chevalier P, Pourchon F and Rinaldi N 2010 *Proc. IEEE BCTM (Austin, Texas, USA)* 137
[9] Hasnaoui I, Pottrain A, Gloria D, Chevalier P, Avramovic V and Gaquière C 2012 *IEEE Electron Device Letters* **33** 1762
[10] Dawson D E, Gupta A K and Salib M L 1992 *IEEE Trans. Electron Devices* **39** 2235
[11] Liu W and Khatibzadeh A 1994 *IEEE Trans. Electron Devices* **41** 1698
[12] Nenadović N, d'Alessandro V, Nanver L K, Tamigi F, Rinaldi N and Slotboom J W, 2004 *IEEE Trans. Electron Devices* **51** 51
[13] d'Alessandro V and Rinaldi N 2002 *Solid-State Electronics* **46** 487

Electro-Thermal Analysis and Optimisation of Edge Termination of Power Diode Supported by 2-D/3-D Numerical Modelling and Simulation

P Príbytný, D Donoval, A Chvála, J Marek, and M Molnár
Institute of Electronics and Photonics, Faculty of Electrical Engineering and Information Technology, Slovak University of Technology, Ilkovičova 3, Bratislva 812 19 Slovakia

E-mail: patrik.pribytny@stuba.sk

Abstract. Numerical modelling and simulation provide an efficient tool for analysis and optimization of device structure design. In this paper we present the analysis and the geometry optimization of the power module with high power pin diode structure supported by the advanced 2-D/3-D mixed-mode electro-thermal device simulation. The structure under investigation is $P^{+}NN^{+}$ power diode device designed for high reverse voltages and very high forward currents, with a maximum forward surge current up to 2.7 kA.

1. Introduction

Various physical phenomena in semiconductor devices are very complex and, depending on applications, are described by partial differential equations of different level of complexity [1]. The parameters of electro-physical models and the boundary conditions of used equations can be very complex. They depend on microscopic physics, the structure of the device, it's actual temperature, and applied biases. Device simulator SDevice [2] allows to simulate electrical characteristics for selected combinations of transport equations and physical models. These models provide the potential to simulate a wide spectrum of semiconductor devices. Many parameters of power semiconductor devices are temperature dependent. Their thermal behaviour contributes to the overall performance of the system. Thermodynamic simulation allows analysis of the influence of structure edge termination profile on the electrical properties of analysed diode. Lowering the diode temperature by structure edge termination shape will allow withstanding the higher current pulses and suppressing the damage of the analysed structure by thermal breakdown. Electro-thermal simulation provides an overview of the chip power and heat dissipation and corresponding temperature distribution for various pulses of load currents [3]. 3-D numerical modelling and simulation using well calibrated parameters provide an efficient tool for analysis and optimization of the device structure design, prediction of its properties, and electro-physical interpretation of obtained data as well.

2. Problem description

Maximum peak forward current is the maximum current which can diode sustain in forward bias mode. Its value is limited by the diode junction thermal resistance and thermal capacity and it is usually much higher than the average operational current due to thermal inertia. It is based on a fact that it takes a non-negligible finite time interval for the diode to reach the maximum temperature for a given current. Ideally, this value should be infinity [4]. This current peak can generate intensive heat

(Joule losses) due to the thermal resistance of metallization and passivation layers. The heat dissipation into the individual layers and their interfaces may be insufficient and it can lead to uncontrolled temperature increase which finally destroys the diode [5]. Therefore, it is necessary to find the critical regions and to optimize the device structure. 2/3-D numerical electro-thermal modelling and simulation provide a very effective solution of this problem. In a mixed mode electro-thermal simulation we combine the electro-thermal simulation of diode chip and analysis of the chip packages from a thermal budget management perspective. The framework required for such analysis is developed and some examples are presented to illustrate its role and effectiveness. The electro-thermal simulations were performed by SDevice tool. There is a possibility to simulate the whole spectrum of semiconductor devices, from high-power to submicron devices. The thermodynamic (or non-isothermal) model [6] extends the drift-diffusion approach to account for electro-thermal effects, under the assumption that the charge carriers are in thermal equilibrium with the lattice. The thermodynamic model is defined by the Poisson equation, current continuity equations, and the lattice heat flow equation (equation 2). The relations generalized to include the temperature gradient as a driving term:

$$J_{n,p} = -nq\mu_{n,p}(\nabla\phi_{n,p} + P_{n,p}\nabla T) \tag{1}$$

where J_n is the electron current density, J_p is the hole current density, μ_n and μ_p are the electron and the hole mobilities, Φ_n and Φ_p are the electron and the hole quasi-Fermi potentials, P_n and P_p are the absolute thermoelectric powers [7]. To calculate the temperature distribution inside the device due to self-heating, the following equation is solved:

$$\begin{aligned} c\frac{\partial T}{\partial t} - \nabla\cdot\kappa\nabla T = -\nabla\cdot\left[(P_nT+\phi_n)\vec{J}_n + (P_pT+\phi_p)\vec{J}_p\right] - \left(E_C + \frac{3}{2}k_BT\right)\nabla\cdot\vec{J}_n \\ -\left(E_V - \frac{3}{2}k_BT\right)\nabla\cdot\vec{J}_p + qR(E_C - E_V + 3k_BT) \end{aligned} \tag{2}$$

where κ is the thermal conductivity and c is the lattice heat capacity. E_C and E_V are the conduction and the valence band energies, respectively, and R is the recombination rate. Simulator performed the analysis for small part of the investigated device. By default, SDevice assumes a 'thickness' (effective gate width along the z-axis) of 1 μm. Area Factor is a multiplier for the electrode currents and charges.

3. Modelling and simulation

The structure under investigation is the P+NN+ power diode designed for the reverse voltages up to 1800 V and the high forward currents up to 100 A with a maximum forward surge current up to 2,7 kA. Due to the large diode dimensions (9,8 × 9,8 mm) and approximate symmetry of the device only a small part of the structure marked by rectangle in figure 1a was used for simulation. The cathode was created by diffusion of a highly doped n-type layer ($N_d = 2\times10^{20}$ cm^{-3}) into the highly resistive (63 Ωcm) n-type substrate with thickness of 350 μm. The anode was formed simultaneously by highly doped p-type layer ($N_a = 2\times10^{20}$ cm^{-3}) on the opposite side of substrate with a pn junction located 290 μm from cathode. To achieve a good agreement between measured and simulated breakdown voltages we need to calibrate the electro-physical models. For simulation of *I-V* characteristics we used University of Bologna model (UniBo) [8] with default values of carrier lifetimes for generation-recombination mechanisms equal to μ_{max} *(electrons)* = 1×10^{-5} s, μ_{max} *(holes)* = 3×10^{-6} s and corresponding values of parameters characterizing Trap Assisted Auger recombination (TAA) $c_n^{TAA} = c_p^{TAA} = 1\times10^{-12}$ cm^{3}s^{-1} [9, 10].

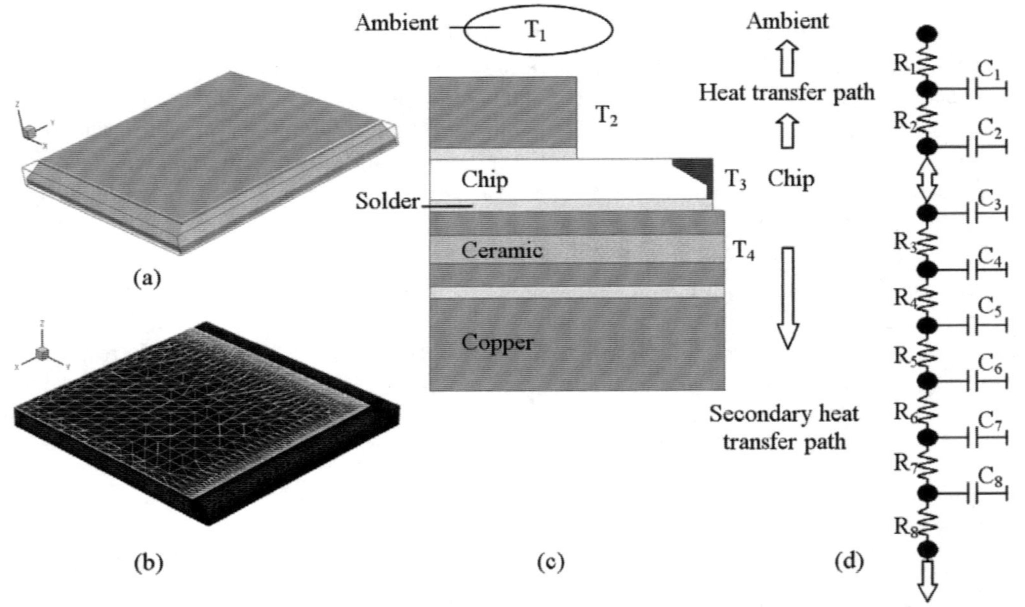

Figure 1. (a) Diode chip without metallization and passivation layer. (b) Structure with passivation layer and grid without package. (c) Cross-sectional view of a chip in package. (d) Thermal RC network.

The best agreement with experimental data we found using τ_{max} *(electrons)* $= 3\times10^{-3}$ *s*, τ_{max} *(holes)* $= 5\times10^{-4}$ *s* and $c_{pn}^{TAA} = 1{,}3\times10^{-12}$ cm^{3}s^{-1} which corresponds to the TAA lifetimes dependent on the carrier densities as follows:

$$\frac{1}{\tau_{n,p}^{TAA}} = c_{p,n}^{TAA}(n+p) \quad (3)$$

The doping dependence of the Shockley-Read-Hall (SRH) recombination lifetimes was modeled in SDevice by the Scharfetter relation.

The experimental characteristics are depicted in range which corresponds to the main application area of the investigated power devices. We achieved a very nice agreement between experimental data and simulations using the calibrated values of parameters. . The simulated value of $V_{BR}^{sim} = 2161$ V nicely coincides with that obtained from experimental measurements $V_{BR}^{exp} = 2187$ V within experimental error (figure 2).

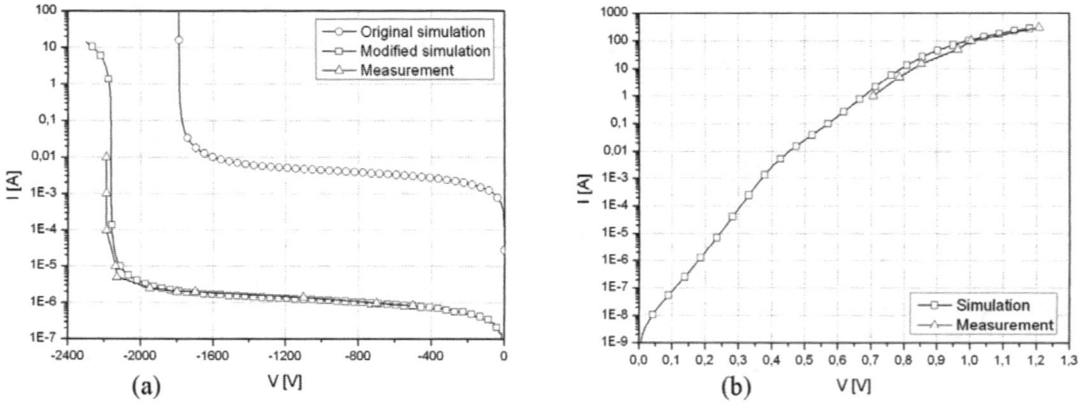

Figure 2. Simulated and measured I–V characteristics. (a) Reverse and (b) forward.

4. Experimental results

As the correct thermal modeling plays a significant role in the combined electro-thermal modeling and simulation of power devices the additional metallization and passivation layers were added to perform more realistic heat dissipation from the diode structure. Figure 4 shows a typical chip in package with a simplified thermal resistance network model. For the short current pulses up to 10 ms we can assume, that the primary heat transfer path through the interface material, spreader and heat sink to the ambient on the top of the structure is of sufficiently low thermal resistance, so that any heat dissipation through the secondary path can be neglected due to high thermal resistance of ceramic block [11, 12]. The heat path from the chip to the back of the lead frame or to a heat sink can be modeled, for example, using the transmission line equivalent circuit diagram. In this case, the current source represents the power dissipation (heat flow) occurring in the chip in the thermal equivalent. Analogous to the electrical transmission line, thermal resistance and thermal capacitance (figure 1d) per unit length need to be considered for exact characterization of the thermal properties of individual components. However, to simplify the description, it is helpful and permissible to combine single volume elements in lumped equivalent elements (R_{th}, C_{th}). Both contact and component resistances are included in these quantities. High thermal resistance of passivation layers and their interfaces prevent the heat dissipation from the chip, which subsequently can destroy the diode. The simple diode structure (without passivation layers and metallization) sustain maximum peak current I =1100 A in the dynamic simulation, while the measured maximum peak current was I =2700 A. Electro-physical model of the diode with additional layers was used for the simulations and for the analysis of internal properties of the diode structure during the peak current pulse I =3100 A.

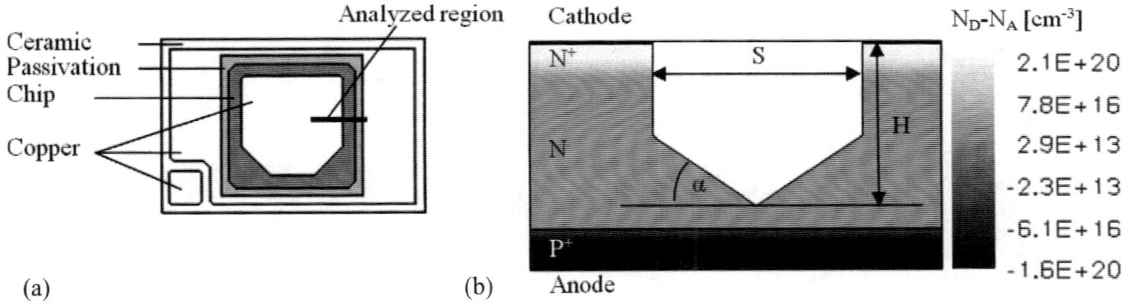

Figure 3. (a) Structure layout, (b) Semiconductor cross section: S, H, α – parameters of edge termination.

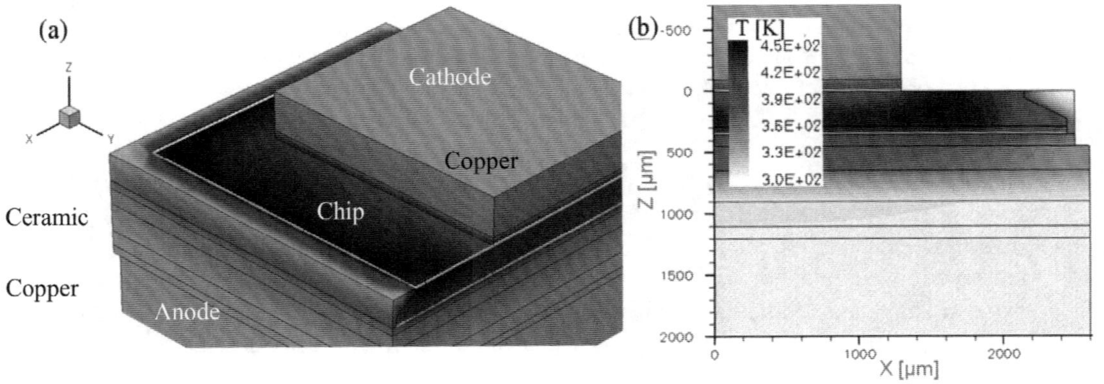

Figure 4. Temperature distribution (a) 3-D profile, (b) 2-D simulation cut during 3100 A inside structure int the time of 7.5×10^{-3} s.

We focused on structures with geometry varies. Several modifications of the diode edge termination using different shapes (angle α) height (parameter H) and width (parameter S) of side wall were designed (figure 3). Measured samples are structures 1-6 marked in figure 5 c, d. The combinations of different simulation parameters influence the diode structure breakdown voltage and distribution of the current density at the peripheral regions of structure as well. Figure 5 shows the dependence of the maximum temperature on the width, depth and angle of edge termination profile with angles 30° and 60°. Characteristics were made for 3100 A stress and in the time of current pulse 7.5×10^{-3} s. All simulations were performed at the same conditions. We observed, that increasing the angle of edge termination, the maximum temperature decreasing. Increasing the width and depth of the edge termination profile, the temperature rises. The lowest temperature was observed for a structure with the angle of 60°. While for small angles (up to 30°) minimum temperature of 430 K was obtained, for structure with the angle of 60° the temperature reaches its maximum of 429 K from all simulations.

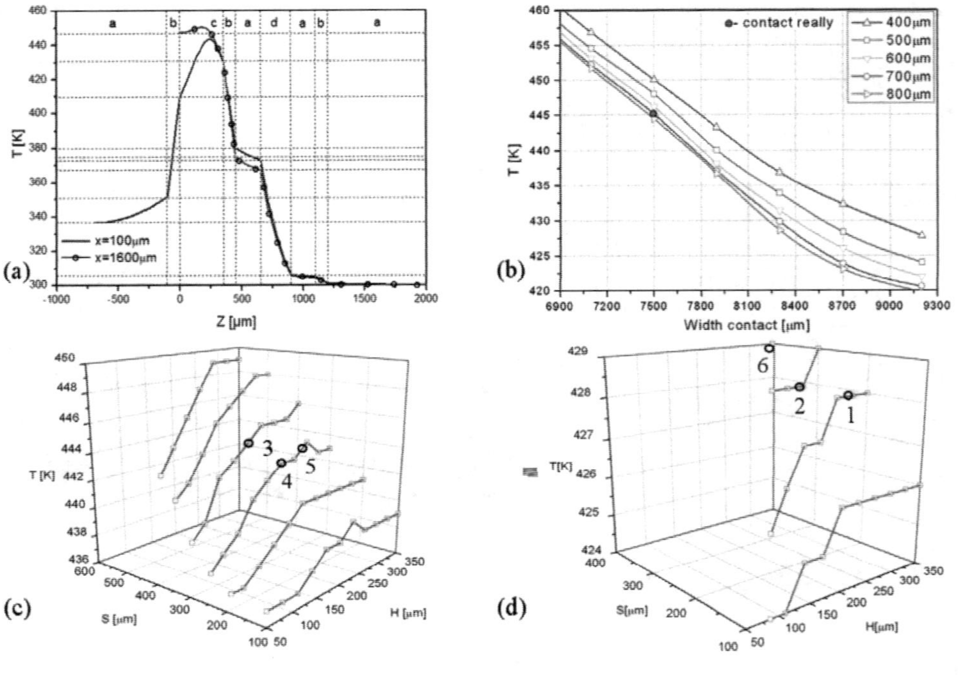

Figure 5. (a) Temperature in 2-D simulation cut, (b) maximum temperature dependence on width and depth of cathode thickness. And max. temperature dependence on height, width and angles (c) 30°, (d) 60° of side wall fewer than 3100 A stresses in the time of 7.5×10^{-3} s.

5. Conclusions

The 2-D numerical device modelling and simulations were applied to the study of electro-thermal behaviour and reliability of analysed device. The simulation of static breakdown voltage V_{BR} was used for the calibration of model parameters. Very good agreement between experimental and simulation results were obtained. Simulation employing the thermodynamic transport model allows analysis of the influence of structure edge termination profile on the electrical properties of analysed diode. The structure with angle 60° exhibits the best results withstanding the current stresses. Lowering the diode temperature by structure edge termination shape will allow withstanding the higher current pulses and suppressing the damage of the analysed structure by thermal breakdown. Finally, good agreement between simulations, measurements and theoretically calculated critical temperatures for diode structure confirm the validity of our approach very well.

6. References

[1] Doering R 2008 *Semiconductor Manufacturing Technology* (Taylor & Francis Group, LLC)
[2] Synopsys, user manual, Version F-2011.09, September 2011
[3] Bagnoli P 1998 *Thermal resistance analysis by induced transient (TRAIT) method for power electronic devices thermal characterization* (IEEE Trans Power Electron)
[4] Hunger T 2007 *Numerical and experimental study on surge current limitations of wire bonded power diodes* (Nuremberg PCIM)
[5] Schroder K 2006 *Semiconductor material and device characterization* (New Jersey: John Wiley & Sons, Inc.)
[6] Wachutka G 1989 *An extended thermodynamic model for the simultaneous simulation of the thermal and electrical behavior of semiconductor devices* (NASECODE, Boole Press Ltd.) 409–14.
[7] Kerr M 2002 General parameterization of Auger recombination in crystalline silicon *Journal of Applied Physics* **91**
[8] Sdevice, user manual, Version F-2011.09, September 2011.
[9] Heinze B 2008 *Effects of metallisation and bondfeets in 3.3 kV free-wheeling diodes at surge current conditions* (ISPS'08 Prague) 223–8
[10] Kerr M 2002 General parameterization of Auger recombination in crystalline silicon *Journal of Applied Physics* **91**
[11] Pluschke N 2012 *Novel packaging technology for power modules* (ISIE, IEEE) 420-424
[12] Jinlei M 2012 *A method for thermal resistance calculation of power module considering temperature effect on thermal conductivity* (IPEMC'12) 245-248

Acknowledgments

This work has been done with support of grant VEGA 1/0866/11.

A Circuital Model of Switching Behaviour of 4H-SiC p^{+}-n-n^{+} Diodes Valid at any Current and Temperature

S Bellone[1], F Della Corte[2], L Di Benedetto[1] and G D Licciardo[1]
[1] University of Salerno, D. I. In.,
Via Giovanni Paolo II, 132, Fisciano Salerno, IT
[2] Mediterranea University of Reggio Calabria, D.I.M.E.T.,
Via Graziella Feo di Vito, Reggio Calabria, IT

E-mail: sbellone@unisa.it

Abstract. A circuital model of 4H-SiC p^{+}-n-n^{+} diodes is presented, which is able to describe the switching behaviour of the devices in a wide range of current, voltage and temperature, at an arbitrary instant, with comparable accuracy of numerical simulations. The model has been analytically derived under generic conditions and is capable to calculate also the dynamic spatial distribution of minority carriers in the epitaxial layer. The accuracy of the model is shown by comparison with numerical simulations and experimental measurements.

1. Introduction

Contrarily to unipolar devices in Silicon Carbide, which are already commercially available [1], bipolar devices are still under investigation because of the stronger sensitivity of their electrical behaviour to material properties [2,3] and fabrication process [4,5]. In this context, the temperature effects on the electrical behaviour of bipolar devices has not been sufficiently modelled so far, although a number of relevant dependencies have been observed. In the present paper, starting from previous works of the authors [6,7], where the combined effects of the apparent bandgap narrowing and incomplete ionization of doping on the steady-state behaviour of 4H-SiC *p-i-n* diodes have been modelled, a circuital model of the devices is presented, which is able to describe their switching behaviour in a wide range of current, voltage and temperature. Although numerical simulation tools have reached an impressive degree of complexity in exploring the electrical characteristics of devices, physics-based models continue to assume an important role for a meaningful interpretations of the experimental behavior of devices and the development of compact models for circuit simulation programs. Currently, a large part of analytical models used for the static description of SiC diodes [8-10] are based on the Herlet's model [11], which was originally proposed for very high current operation of silicon *p-i-n* diodes, while the dynamic models are based on charge control equations [12]; both leave a large space in the literature to propose models capable of describing the characteristics of SiC diodes from the lowest to the highest injection regimes at different temperatures. The proposed model is also capable to calculate the dynamic spatial distribution of minority carriers in the base of diodes with an accuracy comparable to numerical simulations, avoiding any fitting parameters, which extends the model validity to wider domains if compared to compact models, and makes it a useful instrument for the study of bipolar devices in 4H-SiC and their applications.

2. Circuital Model

The proposed circuital model is schematized in figure 1. The S_W switch positions, *S* and *D*, select the static or dynamic behaviour of the diode. All the circuital components in figure 1 are determined once the minority carrier distribution in the epilayer, *p(x)*, has been calculated as in [13-15] by analytically solving the semiconductor diffusion equations. For brevity the solution is not reported and the authors remand to the above references for more details. In the following, the apex '0' will denote static values of time-varying electrical quantities.

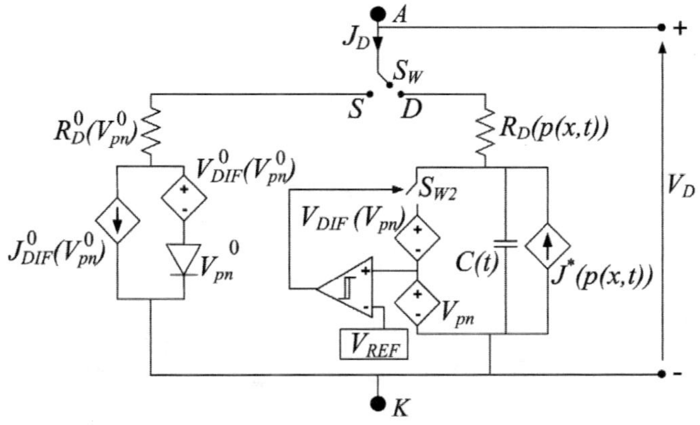

Figure 1. Circuit schematic of the model.

2.1. Static model

The static model [6,7] is composed of two parallel branches. The J_{DIF}^0 current generator accounts for three current components:

$$J_{DIF}^0 = J_{NP^+}^0 + J_{PN^+}^0 + J_{EPI}^0 \tag{1}$$

The first two components of equation (1) describe the minority currents injected in the anode and cathode regions, respectively:

$$J_{NP^+}^0 = qS_{P^+}(T)\,p^0(0^+)\left(1+\frac{p^0(0^+)}{N_{EPI}}\right) \tag{2}$$

and

$$J_{PN^+}^0 = qS_{N^+}(T)\,p^0(W_{EPI}^-)\left(1+\frac{p^0(W_{EPI}^-)}{N_{EPI}}\right) \tag{3}$$

where q is the elementary charge, N_{EPI} is the epilayer doping concentration, S_{P^+} and S_{N^+} are the equivalent temperature-dependent recombination velocities of the P^+ and N^+ terminal regions, respectively, and $p^0(0^+)$ and $p^0(W_{EPI}^-)$ are the hole densities at the epilayer side of the p^+-n and n-n^+ junctions, respectively. The third component of equation (1) describes the recombination current in the epilayer:

$$J_{EPI}^0 = \frac{qL_a}{\tau_a}\frac{\cosh\left[\left(W_{EPI}/L_a\right)-1\right]}{\sinh\left(W_{EPI}/L_a\right)}\left(p^0(0^+)-p^0(W_{EPI}^-)\right) \tag{4}$$

expressed in terms of the temperature-dependent ambipolar lifetime $\tau_a = \tau_p + \tau_n p^0 / \left(p^0 + N_{EPI}\right)$ and ambipolar diffusion length L_a. The other branch accounts for the recombination current in the space charge region, W_{SC},

$$J_{SC}^0 = q \frac{W_{SC}\sqrt{N_{EPI}p^0}}{2\tau_{SC}} \tag{5}$$

where τ_{SC} is the carrier lifetime in the space charge region *(SCR)* and such current density is represented by the diode symbol in figure 1, to which the Dember voltage contribution is in series:

$$V_{DIF}^0 = V_T \ln\left[\frac{(b+1)p^0(0^+) + bN_{EPI}}{(b+1)p^0(W_{EPI}^-) + bN_{EPI}}\right]^{\frac{b-1}{b+1}} \tag{6}$$

where $b=\mu_n/\mu_p$ is the electron-hole mobility ratio. The series resistance, R_D^0, accounts for the constant resistances of the terminal regions, including the contacts, and the injection dependence resistance of epilayer [6]. The current–voltage relation of the static model is finally given by

$$V_D^0 = V_{PN}^0 + V_{DIF}^0 + R_D^0\left(J_{SC}^0 + J_{DIF}^0\right) \tag{7}$$

where the p^+-n junction voltage, V_{PN}^0, is calculated by the well-known junction law. In figure 2 the static analytical J_D-V_D curves of a 10µm-thick epilayer are compared with those from Silvaco ATLAS simulations [16] in a wide range of current and temperature values. It is interesting to note that the curves at different temperatures never intersect, which indicates that the diode series resistance presents a negative temperature coefficient [17].

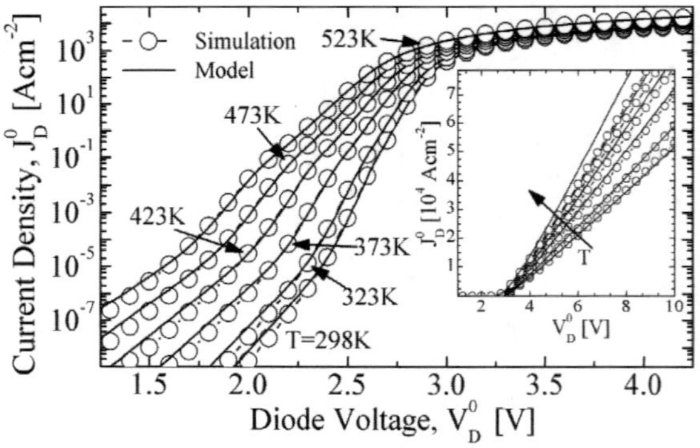

Figure 2. Steady-state J_D-V_D curves of 10-µm thick diode.

2.2. Dynamic model

When the switch S_W in figure 1 is positioned on D, the switch S_{W2} remains closed until the minority carrier density at the epilayer border of the p^+-n junction reduces to N_{EPI}, namely until $V_{PN} > V_{REF} = 2V_T ln\left(N_{EPI} / n_{io}\right)$ [15]. During this *storage* interval ($t<t_S$), the current-voltage characteristic of the diode is determined by the same quantities introduced above for the static model,

Journal of Physics: Conference Series **494** (2014) 012004 doi:10.1088/1742-6596/494/1/012004

expressed this time in terms of the time-varying hole concentration. After the storage ($t>t_S$), S_{W2} opens and the diode turn−off phase is described by the charging of the time-varying junction capacitance

$$C_j = \frac{\varepsilon}{W_{SC}}\left[1+\frac{p(0,t_S)e^{-\frac{t-t_S}{\tau_a}}}{N_{EPI}}\right] \tag{8}$$

by means of the current density determined by the difference between the total diode current and J^* given by $J_p(W_{SC},t)+J_{pC}(t)$. The first term represents the hole density variation in the epilayer due to W_{SC} widening, given by:

$$J_p(W_{SC},t)=qD_p\frac{p(x_{MAX}(t_S),t_S)-p(W_{SC},t_S)}{x_{MAX}(t_S)-W_{SC}(t_S)}e^{-\frac{t-t_S}{\tau_{p,EQ}}} \tag{9}$$

where x_{MAX} is the epilayer abscissa where the hole density assumes its maximum value p_{MAX}, and $\tau_{p,EQ}$ has been calculated in [18]. The second term represents the current flowing through the substrate (3), approximated with the cathode current related to the peak value as [6]:

$$J_{pC}(t) \cong qS_{N^+}p_{MAX}(t)\left(1+\frac{p_{MAX}(t)}{N_{EPI}}\right) \tag{10}$$

The accuracy of the dynamic model is shown in figure 3 by comparing current and voltage transients, at different temperatures, with numerical simulations of the same *10−μm* thick diode of figure 2, when it is forward biased at $100Acm^{-2}$ and then switched on a resistive load with a power supply of $-20V$.

The proposed model is isothermal and does not account for self-heating effects. Its validity is essentially limited by the temperature dependences of the transport parameters involved in the model, among which the mobility is particularly critical. The validity of the implemented mobility model [6] has been experimentally verified for silicon carbide in [19] up to about 600K, while in [20] an electro-thermal model of a lateral SiC JFET, using the same mobility equations, has been verified up to 300°C by simulation comparisons. These results suggest that the proposed model can be successfully employed up to about 600K. In this range of temperature, the model reveals an useful instrument for the dimensioning of diodes in the design process of switching circuits, like power converters, where (4)-(6) and (8)-(10) allow one to accurately define the *p-n* junction structure and the epilayer characteristics (thickness, doping and ambipolar carrier lifetime) to optimize the switching velocity and the dynamic power losses. Similarly for communication systems, the model can be used to calibrate the series resistance of diodes for RF power switches and attenuators to minimize the insertion losses.

Journal of Physics: Conference Series **494** (2014) 012004 doi:10.1088/1742-6596/494/1/012004

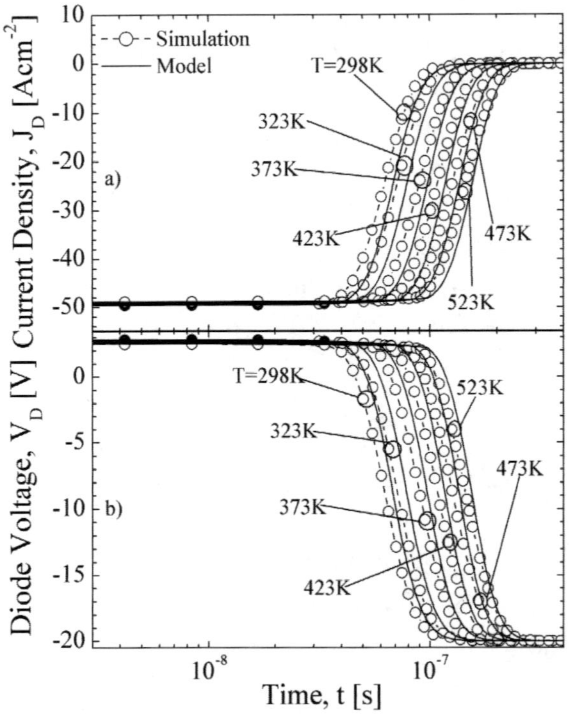

Figure 3. Current (a) and voltage (b) transients of 10-μm thick diode.

In figure 4, the capability of the model to describe also the internal quantity distribution is shown by comparing the hole distribution transient with numerical simulation results, sampled at five instants of time and at three temperatures. Finally, a comparison of the model with simulations and experimental results taken on a *25 μm*–thick diode are reported in figure 5 at two temperatures. The device under test is a 4H-SiC *p-i-n* vertical diode, whose anode contact has a circular geometry with a *350 μm* diameter. The $3\times10^{15}\,cm^{-3}$ doped base region is epitaxially grown on a *300 μm* thick and $5\times10^{19}\,cm^{-3}$ nitrogen-doped substrate. The anode region is Al implanted at a depth of *1.3μm* with a peak doping of $6\times10^{19}\,cm^{-3}$. Deposition of Al/Ti and of Ni realizes the ohmic contacts for the anode and cathode regions, respectively. The switching curves are obtained by imposing a forward and reverse current densities of $60\ Acm^{-2}$ and $115\ Acm^{-2}$ with a HP 8114A Pulse Generator and are acquired with Tektronix DPO7254 digital oscilloscope.

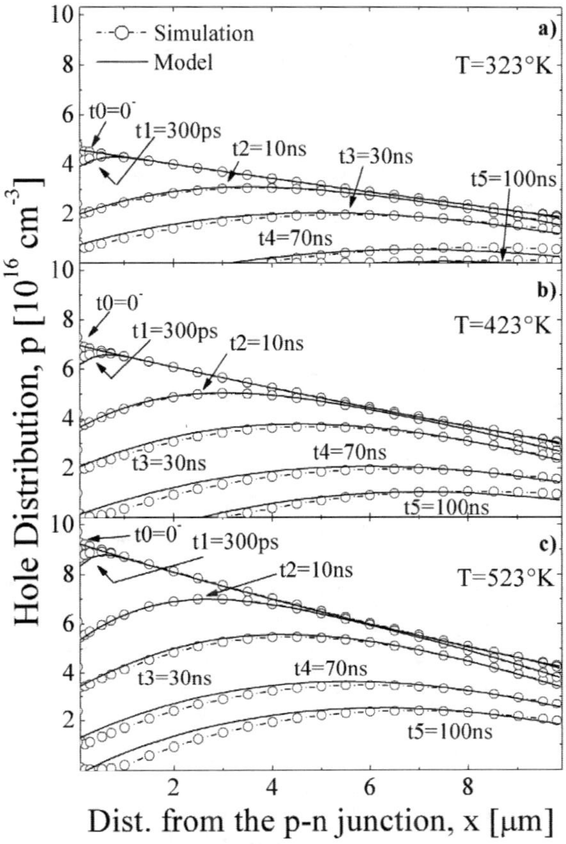

Figure 4. Transients of hole distribution in the same switching conditions of figure 3.

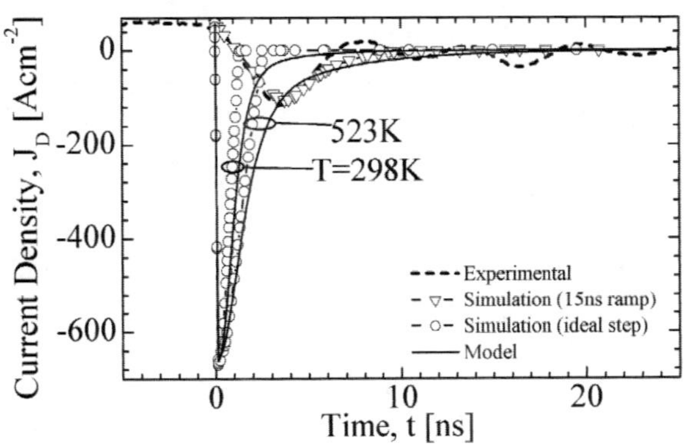

Figure 5. Comparisons among experimental measurement, numerical simulations and model results of a 25-μm thick diode [9].

3. Conclusion

A circuital model of the static and switching behavior of 4H-SiC *p-i-n* diodes has been proposed, which is able to describe the diode *I-V* characteristics at the terminals as well as the spatial-temporal profiles of physical quantities, like hole concentration and electric field. Numerical simulations and measurements have been compared with the model to show its accuracy to describe the diode electric behavior at different temperatures. It can be conveniently used to design 4H-SiC *p-i-n* diodes for power and RF power applications.

4. References

[1] Cree Inc. (http://www.cree.com).
[2] Bellone S Di Benedetto L 2014 A model of the I_D-V_{GS} characteristics of normally-off 4H-SiC bipolar JFETs *IEEE Trans. on Power Electron.* **1** 29 514 – 521
[3] Bellone S Di Benedetto Licciardo G D 2012 A Quasi One–Dimensional Model of the Potential Barrier and Carrier Density in the Channel of Si and 4H–SiC BSIT's *IEEE Trans. on Electron Devices* **59** 9 2546-2549
[4] Buono B Ghandi R Domeij M Malm G B Zetterling C-M Östling M 2010 Modeling and characterization of current gain versus temperature in 4H-SiC power BJTs *IEEE Trans. on Electr. Dev.* **57** 3 704-711
[5] Nipoti R Di Benedetto L Albonetti C Bellone S 2012 Al^+ Implanted Anode for 4H–SiC p–i–n Diodes *ECS Tans. On GaN and SiC Power Tech.*, 50 3 391–397.
[6] Bellone S Della Corte F G Albanese L F Pezzimenti F 2011 An analytical model of the forward I-V characteristics of 4H-SiC p-i-n diodes valid for a wide range of temperature and current *IEEE Trans. on Power Electron.* **26** 10 2835-2843
[7] Bellone S Della Corte F G Di Benedetto L Albanese L F Licciardo G D 2010 A self–consistent model of the static and switching behavior of 4H-SiC Diodes *IEEE International Semiconductor Conference*
[8] Sheng K Zhang Y Yu L Su M Zhao JH 2009 High-frequency switching of SiC high-voltage LJFET *IEEE Trans. Power Electron.*, **24** 1 271–277
[9] Dyakonova NV Ivanov PA Koslov VA Levinshtein ME Palmour JW Rumyantsev SL Singh R 1999 Steady-State and transient forward current-voltage characteristics of 4 H Silicon Carbide 5.5.kV diodes at high and superhigh current densities *IEEE Trans. Electron Devices*, **46** 11 2188–2194
[10] Abuishmais I Undeland TM 2010 Simplified models of forward conduction for SiC power pin and Schottky diodeswith temperature dependency in *Proc. 5th IET Int. Conf. Power Electron., Mach. Drives* 1–6.
[11] Herlet A 1968 The forward characteristic of silicon power rectifiers at high current densities *Solid State Electron.* **11** 717–742
[12] Lauritzen PO Ma CL 1991 A simple diode model with reverse recovery *IEEE Trans. on Power Electron.*, **6** 2 188-191
[13] Bellone S Albanese L F Licciardo G D 2009 A self-consistent model of the OCVD behavior of Si and 4H-SiC p^+-n-n^+ diodes *IEEE Trans. on Electr. Dev.* **56** 12 2902-2910
[14] Bellone S Albanese L F Licciardo G D 2009 Limitations of the open-circuit voltage decay technique applied to 4H-SiC diodes *IEEE International Semiconductor Conference*
[15] Bellone S Della Corte F G Di Benedetto L and Licciardo G D 2012 An analytical model of the switching behavior of 4H-SiC p+-n-n+ diodes from arbitrary injection conditions *IEEE Trans. on Power Electronics* **27** 3 1641-1652
[16] ATLAS User's Manual 2005 SILVACO Int. Santa Clara, CA, USA, Ver.5.10R
[17] Satoh M, Nagata S, Nakamura T, Doi H and Shibagaki M 2009 Doping Level Dependence of Electrical Properties for p+n 4H-SiC Diode formed by Ion Implantation *Materials Science Forum* **615-617**, 679-682.

[18] Bellone S, Neitzert HC and Licciardo GD 2004, Modeling and Characterization of the OCVD Response at an Arbitrary Time and Injection Level *Solid State Electronics* **48** 7 1127-1131.

[19] Roschke M Schwierz F 2001 Electron mobility models for 4 H, 6 H, and 3 C SiC *IEEE Trans. Electron Devices*, **48** 7 1442–1447

[20] Zhong X Wang T Guo Q Sheng K 2011 High Temperature Physical Modeling and Verification of 4H-SiC Lateral JFET Device in *IEEE 2nd International Conference on Artificial Intelligence, Management Science and Electronic Commerce (AIMSEC*), 4015-4018

Acknowledgments
The authors wish to thank Dr. Roberta Nipoti from CNR-IMM in Bologna, Italy, for manufacturing the 4H-SiC diodes.

Hall Mobility Maps for 4H-Silicon Carbide by Monte Carlo Simulations

J Woźny, Z Lisik and J Podgórski

Lodz University of Technology, Department of Semiconductor and Optoelectronic Devices, Wolczanska 211/215, 90-924 Lodz, Poland

E-mail: janusz.wozny@p.lodz.pl

Abstract. The Monte Carlo Single Particle approach was used to analyze electron transport in 4H-SiC taking into account the influence of the magnetic field. Within the numerical approach it was possible to evaluate electron Hall mobility and the Hall factor for the wide range of donor concentrations and temperatures varying from 300 K up to 700 K.

1. Motivation

The Silicon Carbide (SiC) is the wide band gap semiconductor which technology is under continuous development. Due to properties of its hexagonal politypes (6H-SiC, 4H-SiC) the SiC is dedicated to high temperature-, high frequency- and power-electronics. During last years, mostly due to technological problems eq. degradation of transport parameters during SiC processing a lot of research groups directed their attention to GaN. However it appeared that due to lack of good quality GaN monocrystals some attention was turned back to SiC in the field of high voltage devices [1], Thus it seems that currently SiC is the only material appropriate for high power electronics [1]. Still there is lack of complete database of macroscopic parameters like mobility where the dependence on a wide range of environmental parameters (temperature, dopant concentration, crystallographic orientation) would be included. Measurements of SiC drift mobility μ_D [2] required special geometries and were done for very limited cases. More popular measurements of mobility utilize Van Der Pauw approach [3], but then the Hall mobility μ_H is measured which is different to the drift mobility. To find missing values of macroscopic transport parameters the computational physics can be utilized. The method selected should be based on basic physical phenomena valid for wide range of external conditions but still it has to computationally effective. These requirements are fulfilled by the Single Particle Monte Carlo Method (SPMC) [4] used widely to investigate charge transport in semiconductors [5]. With SPMC applied to SiC with magnetic field included one can calculate both μ_H, μ_D and its ratio, the Hall factor r_H, for wide range of doping concentration and temperatures thus it also allows for recalculation of each type of mobility when the other is obtained by mean of other approach eq. from measurements. Within this paper we would like to present results for electron mobility calculation since electron unipolar SiC high power devices are being currently developed due to nearly one order higher mobility of electrons compared to holes [6,12].

2. Model

The Single Particle Monte Carlo approach was used to analyze the electron transport in 4H-SiC. The models utilized are described in [9]. Following models were used:

- single valley analytical non-parabolic ellipsoidal band-structure

- analytical description of the following scatterings: ionized impurities, acoustic phonon, zero order non-polar optical phonon, first order nonpolar optical phonon, polar optical phonon, impact ionization
- the magnetic and electric field were simply taken into account using the following equations

$$\Delta \boldsymbol{k}_E = -\frac{e\boldsymbol{E}}{\hbar}\tau_{fl} \tag{1}$$

$$\Delta \boldsymbol{k}_B = -\frac{\boldsymbol{v}\times\boldsymbol{B}}{\hbar}\tau_{fl} \tag{2}$$

where $\Delta \boldsymbol{k}_{(B,E)}$ – wave vector change due to presence of electric $\boldsymbol{E}$ and magnetic $\boldsymbol{B}$ field, τ_{fl} - free flight time, $\hbar$ - reduced Planck constant, e – electron charge.

The use of analytical band-structure is justified since the mobilities and Hall factor should be evaluated for low range of energies to obtain correct value of r_H which is calculated for constant value of μ_H and μ_D. The use of Monte Carlo method was previously considered inadequate [10] for evaluation of μ_H due to weak effect of magnetic field in comparison to the thermal noise. However we found that the Monte Carlo method can be used to calculate the effect of magnetic field for the temperature range up to 700 K and for wide range of donors concentration. The magnetic field $\boldsymbol{B}$ we used during simulation was equal up to 2 T which is a value used in Van Der Pauw method measurements systems [7]. The electric field $\boldsymbol{E}$ must be lower than 1E+6 V/m to provide linear dependence of drift velocity on E. However one should keep E high enough to reduce the influence of the thermal noise.

3. Results

With the Monte Carlo tool proposed it was possible to obtain 4H-SiC electron Hall mobilities for different orientation of the crystal, figure 1. The average anisotropy ratio between Hall mobilities for different crystal orientation is found to be 0.7, which is close to 0.8 reported in literature for the drift mobility [11].

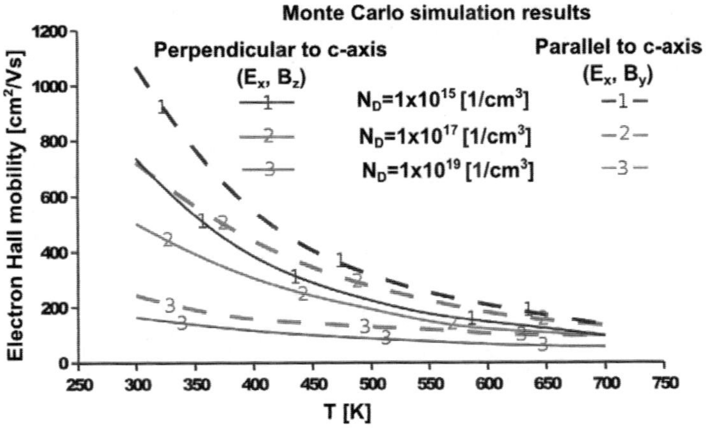

Figure 1. 4H-SiC electron Hall mobility as a function of temperature for different crystal orientation.

One can also use the method to obtain the whole map of the 4H-SiC electron Hall mobilities as a function of temperature and dopant concentration. Such map is presented in figure 2. Similar maps could be obtained from analytical models eq. the Caughey-Thomas (C-T) model [8,11] which also takes into account influence of doping concentration and temperature. Comparison with this model having parameters taken from [6] is depicted in figure 3. For the sake of clarity only contours of such maps are presented. The shape of the mobility maps for both methods is the same. However the values obtained from Caughey-Thomas model are higher in the range of lower temperatures and dopant concentrations. This could be explained by the fact that the Caughey-Thomas model reflects empirical dependence and is based on measurements of limited number of SiC crystal samples. The Caughey-Thomas (or any) empirical model reflects carrier transport in certain set of samples or some average of different measurements done by different groups and do not contain direct information about the physics inside. The proposed Monte Carlo method includes basic physics and is also depended on certain set of parameters. However the advantage of Monte Carlo simulations is that it is based on physical models and can be extended to catch other phenomena of carrier transport in SiC, eg. crystal defects.

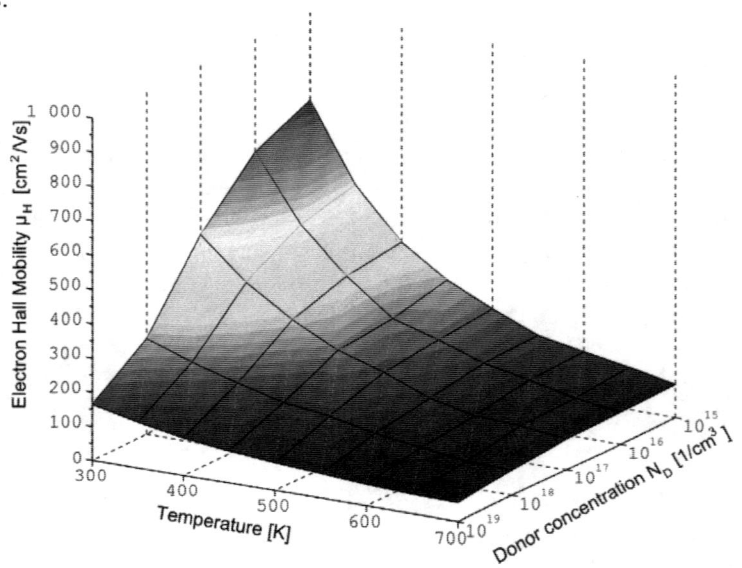

Figure 2. Electron Hall mobility map perpendicular to c-axis.

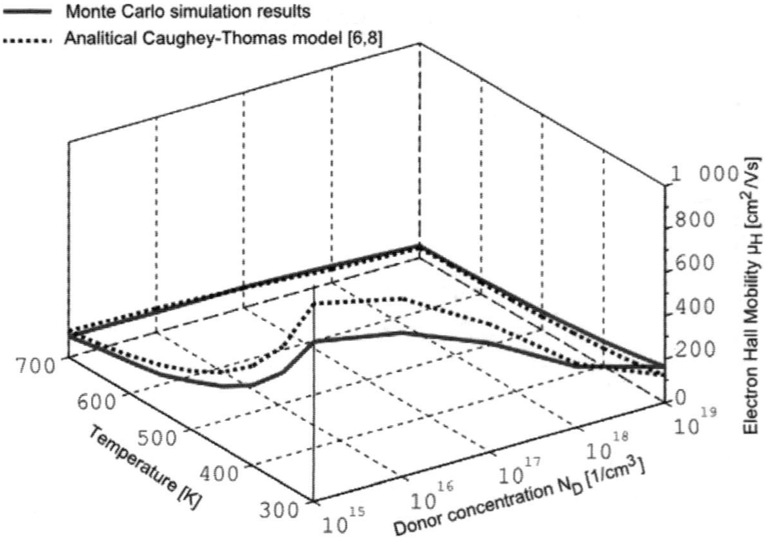

Figure 3. Electron Hall mobility map perpendicular to c-axis compared with Caughey-Thomas model.

Measurement results of the Hall mobilities available in the literature [12,13] are also shown in figure 4

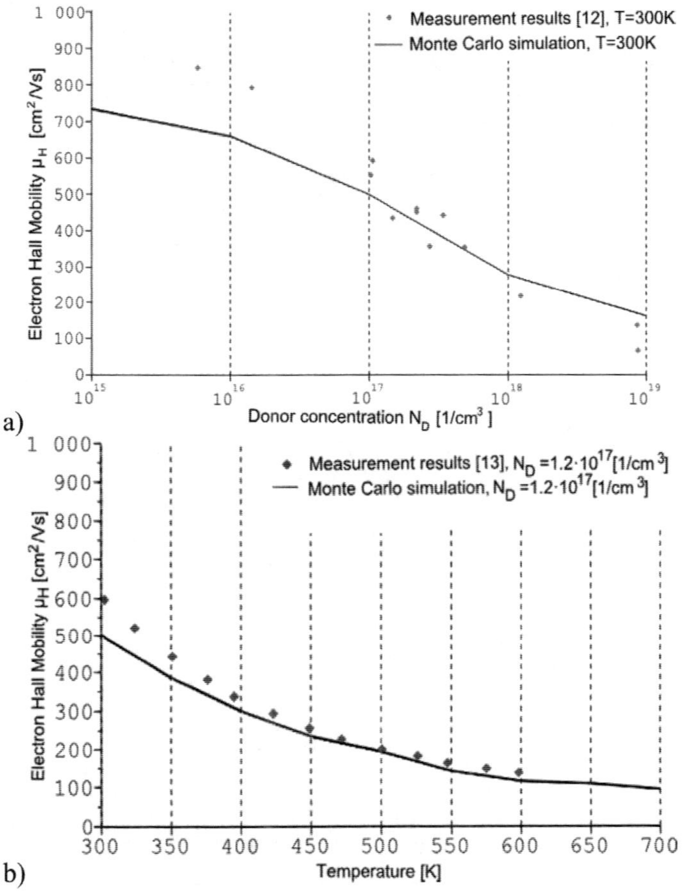

Figure 4. 2D cross-sections of the Hall mobility map from figure 2 compared with measurement results taken from literature a) [12], b) [13].

Comparison presented in figure 4 shows that the calculated Hall mobility are in the range of measured values [12,13]. No perfect matching can be caused by different quality of SiC samples. The SiC crystal quality is continuously improving but it may be still defected during fabrication processes thus comparing simulation data with referenced measurements is difficult. Taking into account crystal defects in simulations is planned in future.

The calculated Hall factor is depicted in figure 5. Having knowledge about this parameter one can recalculate μ_H obtained from measurements to receive proper value of μ_D

Comparison with data available in literature [14,15] (figure 6) shows that obtained r_H agrees with values obtained by other methods. Unfortunately, to the best knowledge of the author, there is no reference data for temperatures higher than 400 K.

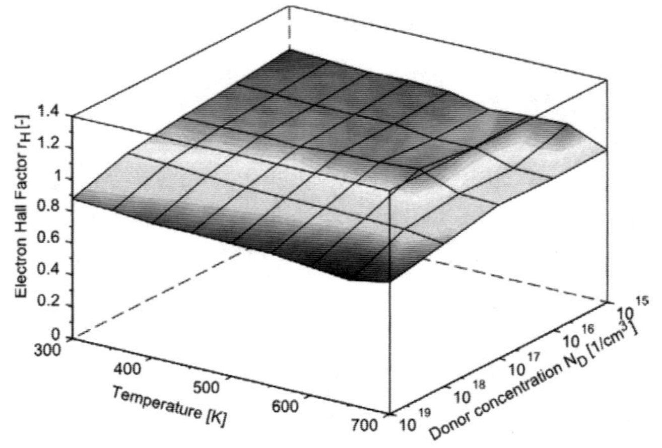

Figure 5. Hall factor r_H perpendicular to c axis as function of the temperature and donor concentration.

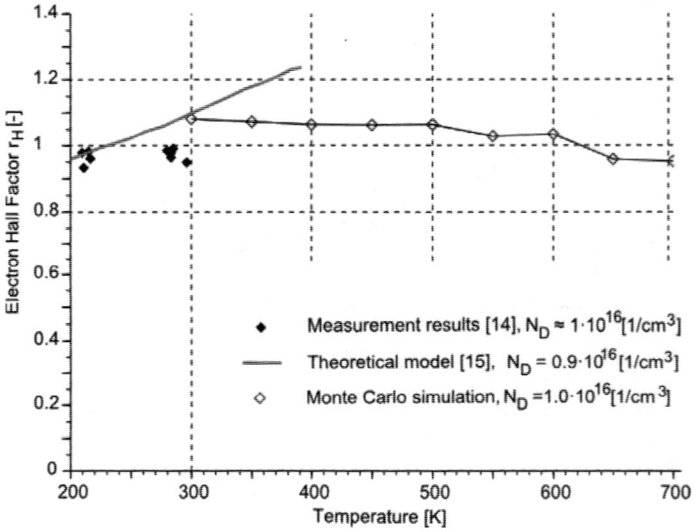

Figure 6. Hall factor r_H perpendicular to c axis compared with available data.

4. Summary

By mean of the Monte Carlo method it is possible to evaluate the Hall electron mobility μ_H and Hall factor r_H for 4H-SiC. Previously the Monte Carlo method was used to evaluate drift mobilities which could not be directly compared with large set of measurements since most of the measurements were done using Van Der Pauw method where the μ_H is obtained. With μ_H obtained from Monte Carlo one can evaluate maps of parameters that can be used in other CAD tools. Moreover direct access to measured and simulated μ_H allows for scattering model corrections which makes possible the creation of unique mobility maps (as function of eq. doping, temperature) for specific 4H SiC wafers.

5. References

[1] Roussel P Will GaN-on-Si displace Si and SiC in power electronics? 2011 *Proc. of the 2011 Int. Conf. on Compound Semiconductor Manufacturing Technology (CS MANTECH)* Palm Springs California USA pp 16-19

[2] Mickevicius R and JH Zhao 1998 Monte Carlo study of electron transport in SiC *J. Appl. Phys.* **83(6)** pp 3161-3167
[3] Van Der Pauw LJ 1958 A method of measuring specific resistivity and Hall effect of discs of arbitrary shape *Philips Research Reports* **13** pp 1–9
[4] Tomizawa K 1993 Numerical Solution of Submicron Semiconductor Devices (Arthech House Boston – London)
[5] Jacoboni C and Paolo L 1989 *The Monte Carlo method for semiconductor device simulation Vol 3* (Springer)
[6] Ayalew T. 2004 *SiC semiconductor devices technology modeling and simulation* (PhD Dissertation TU Wien)
[7] Lake Shore Cryotronic 2013 *Hall Effect Measurement System Catalog*
[8] Caughey D and Thomas R 1967 Carrier Mobilities in Silicon Empirically Related to Doping and Field *Proc IEEE* **52** pp 2192-2193
[9] Donnarumma G Woźny J Lisik Z 2009 Monte Carlo simulation of bulk semiconductors for accurate calculation of drift velocity as a parameter for drift-diffusion hydrodynamic models *Materials Science and Engineering: B* **165(1)** pp 47–49
[10] Ng G Vasileska D and Schroder D G 2009 Calculation of the electron Hall mobility and Hall scattering factor in 6H-SiC *J. Appl. Phys.* **106** 053719
[11] Roschke M and Schwierz F 2001 Electron mobility models for 4H, 6H, and 3C SiC [MESFETs] *IEEE Transactions on Electron Devices* ***48*(7)** pp 1442-1447
[12] Schaffer W *at al* 1994 Conductivity anisotropy in epitaxial 6H and 4H SiC *MRS Proceedings* **339(1)** (Cambridge University Press)
[13] Levinshtein M and Michael Shur 1995 *Handbook series on semiconductor parameters Vol 1,2* (World Scientific)
[14] Rutsch G *at al* 1998 Measurement of the Hall scattering factor in 4H and 6H SiC epilayers from 40 to 290 K and in magnetic fields up to 9 T *J. of Appl. Phys.* 84(4) pp 2062-2064
[15] Iwata H and Itoh K M 2001 Donor and acceptor concentration dependence of the electron Hall mobility and the Hall scattering factor in n-type 4H–and 6H–SiC *J. of Appl. Phys.* **89(11)** pp 6228-6234

Three-Dimensional Electro-Thermal Verilog-A Model of Power MOSFET for Circuit Simulation

A Chvála, D Donoval, J Marek, P Príbytný, M Molnár and M Mikolášek
Institute of Electronics and Photonics, Slovak University of Technology in Bratislava
Ilkovičova 3, 812 19 Bratislava, Slovakia

E-mail: ales.chvala@stuba.sk

Abstract. New original circuit model for the power device based on interactive coupling of electrical and thermal properties is described. The thermal equivalent network for a three-dimensional heat flow is presented. Designed electro-thermal MOSFET model for circuit simulations with distributed properties and three-dimensional thermal equivalent network is used for simulation of multipulse unclamped inductive switching (UIS) test of device robustness. The features and the limitations of the new model are analyzed and presented.

1. Introduction

Today's circuit simulators are standard tools in the development and optimization of electronic systems. However, until now the simulations have been limited to electrical functions because the temperature dependences of parameters of simulation models available today are taken into account at the best by changing the static global temperature. In power electronic systems in particular, temperature is one of the critical parameters due to the non-negligible self-heating effects and the fact that many properties of power semiconductor devices are strongly temperature dependent [1]. The standard equivalent models of the power devices for circuit electro-thermal simulations consist of one-dimensional Cauer or Foster type thermal equivalent networks. The one-dimensional heat flow can be insufficient particularly for large power devices and large power pulses. An inhomogeneous temperature distribution caused by three-dimensional heat flow from the semiconductor chip to the package and cooling assemblies can cause an inhomogeneous distribution of the electric properties of the power device along the whole chip. Therefore, it is necessary to split the structure into several cells with thermal interaction in three dimensions. In this paper we present new electro-thermal model of power MOSFET for circuit simulation, which contains a dynamic link between electrical and thermal components description. Also the thermal equivalent network for a three-dimensional heat flow is shown. The designed electro-thermal model of the analyzed power MOSFET is used for a simulation of UIS test [2].

2. Thermal system description

During a search for an electrical analog model for heat conduction inside structure, we assume that a thermal system can be in general modeled with a discrete element electrical circuit, composed by resistances and capacitances, where the temperatures and the thermal powers are considered as voltages and currents, respectively (figure 1b) [3][4]. Power dissipation of the electrical circuit is determined at all times and a current proportional to the dissipated power $I(t)$ is fed into the thermal equivalent network. The V_{Tj} node voltage represents the junction temperature inside the structure. All non-linear electrical temperature dependent parameters of the power structure are driven by the actual temperature distribution inside the structure [5]. Due to a close relationship with physical reality of the one-dimensional heat flow, the parameters for the *RC* equivalent circuit diagram can be derived directly from equation (1) and (2) [6]. The physical variables are specified in their thermal equivalents by using the geometries, thermal conductivities and thermal capacities of materials, where A is the surface, d_i is the thickness, κ and c are the thermal conductivity and heat capacity of the elements. Then we can define:

$$R_i \approx R_{th\,i} = \frac{d_i}{\kappa \cdot A} \tag{1}$$

$$C_i \approx C_{th\,i} = c \cdot d_i \cdot A \tag{2}$$

The modification of one-dimensional Cauer-type equivalent thermal network to three-dimensional network is shown in figure 1c [7].

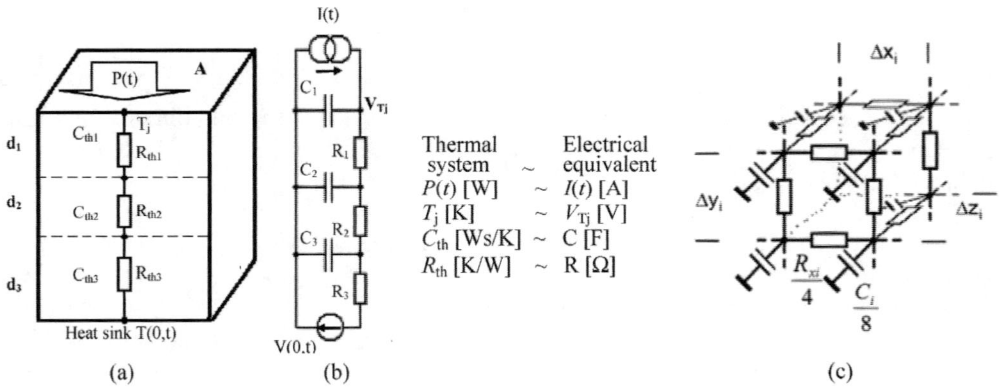

Figure 1. (a) Thermal structure, (b) Cauers electrical equivalent of the thermal system (c) equivalent thermal network for three-dimensional heat flow.

3. Structure description

The structure under investigation is the power super-junction vertical MOSFET [8][9] in DPAK 2 package (figure 2a). The two-dimensional cross section of the structure is shown in figure 2b. The trench gate MOSFET is created on top of the n^{++}-type doped substrate and n^{-}-type doped epitaxial layer. The super-junction trench, created by p-type and n-layers around an air gap, provides charge balance and a high breakdown voltage for low on-resistance. Referring to [10] and device simulation (figure 2c) the most of heat generated during the avalanche operation of UIS conditions is located at the bottom of the air gap epitaxial layer.

Electrical circuit is created by 8 × 8 transistor cells connected in parallel. The power MOSFET is electrically modelled with a SPICE Level 3 built in Verilog-A [11]. The electrical model is defined considering the temperature dependences of the most relevant parameters extracted from the characterization of the structure at different operating temperatures. The temperature dependence of threshold voltage V_{T0}, drain resistivity R_D, body diode D_B, leakage current through leakage resistance R_L, avalanche breakdown voltage V_{BR} and over current/thermal destruction S_{BURN} are implemented in the model [12].

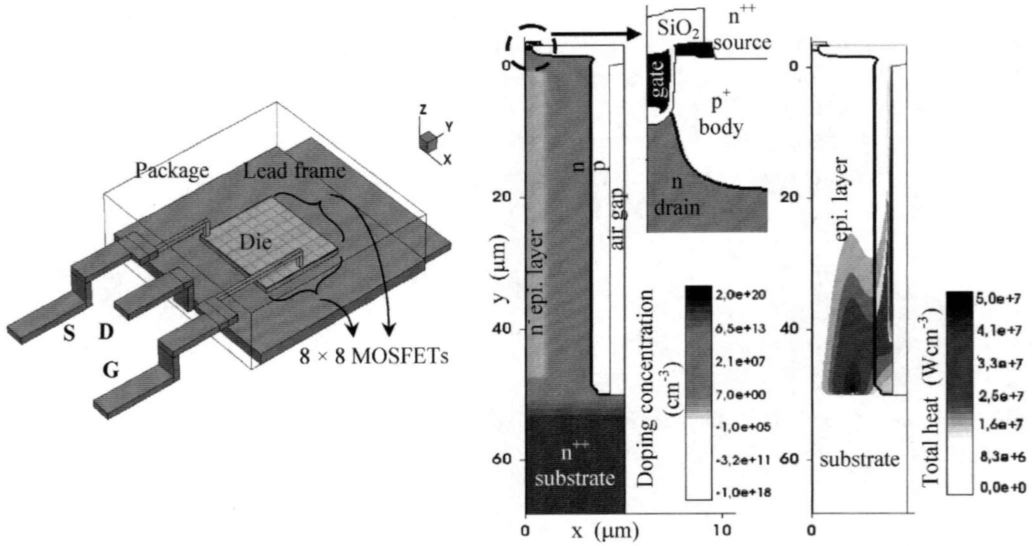

Figure 2. (a) Structure of the MOSFET in DPAK 2 package, (b) two-dimensional cross section of the super-junction MOSFET, (c) total heat distribution inside the structure during UIS breakdown condition.

Figure 3 shows experimental data compared with simulations of transfer characteristics at different temperatures, output characteristics for different gate biases and *C-V* characteristics of the analyzed MOSFET. The three-dimensional thermal equivalent network is created by an RC network allowing three-dimensional heat flow. The three-dimensional mesh is generated automatically by Verilog-A program cycle which allows easy use user defined placement. The thermal material properties, geometry and dimensions are used as the input parameters for mesh creation. Only the lead frame, substrate, epitaxial layer and package are taken into account for the simplicity of the model. The current sources which represent thermal heating are placed between the epitaxial layer and substrate where occurs heat generation during avalanche conditions. The electrical circuit diagram of one transistor cell and view of the tree-dimensional equivalent thermal mesh are shown in figure 4.

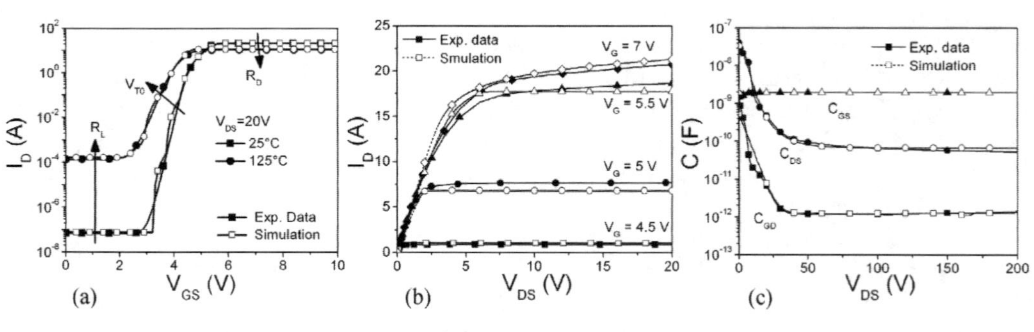

Figure 3. Experimental data and simulation of (a) transfer characteristics at different temperatures, (b) output characteristics and (c) *C-V* characteristics of analyzed MOSFET.

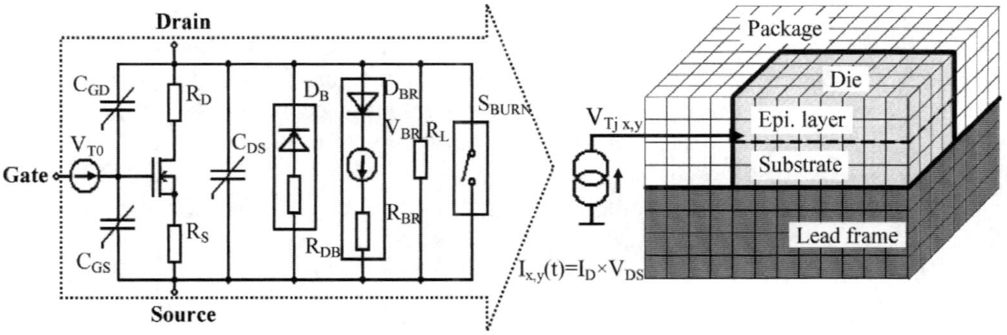

Figure 4. The electrical circuit diagram of one transistor cell (left) and view of the tree-dimensional equivalent thermal mesh (right).

4. Experiment

The designed electro-thermal model of the analyzed power MOSFET has been used to simulate the multipulse UIS test. UIS condition represents the robust test of circuit switching operation for evaluating the ruggedness, which characterizes a device capability to handle high avalanche currents during the applied stress. In case the current flowing through the inductance is quickly turned off, the

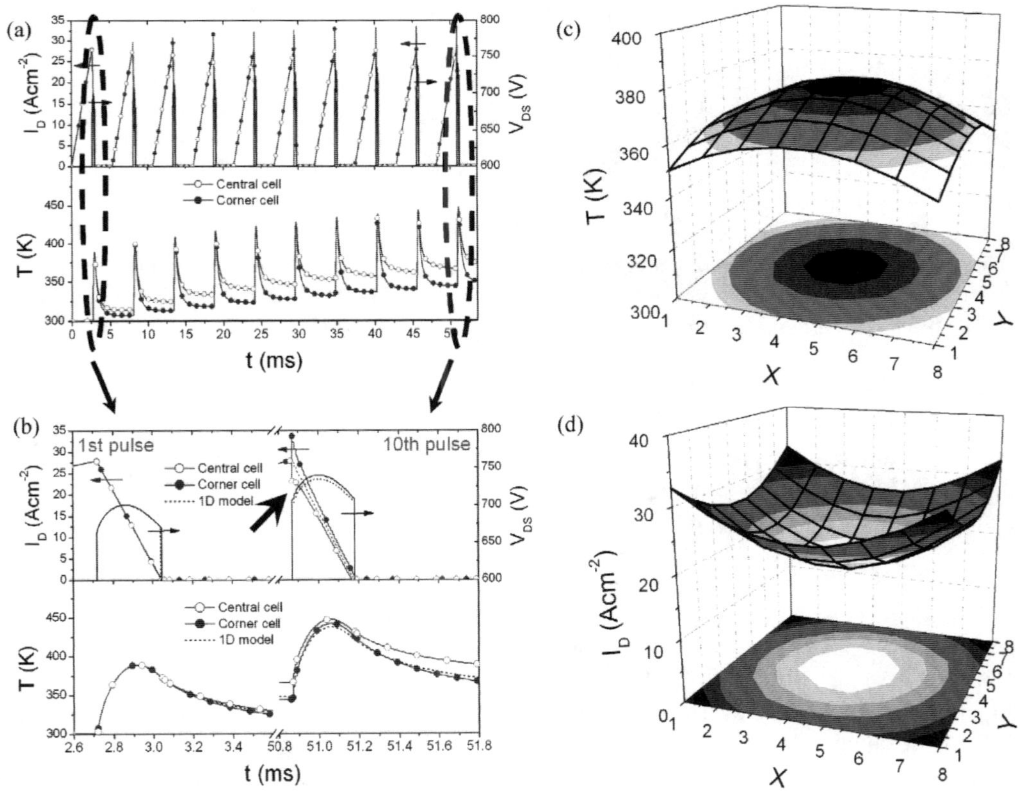

Figure 5. (a) Comparison of the drain current, drain voltage and temperature of the corner and central parts during multipulse UIS test. (b) Comparison of simulation with three-dimensional and one-dimensional thermal network. (c) Temperature and (d) current density distribution at the beginning of the tenth UIS pulse for structure using three-dimensional thermal network.

magnetic field induces a counter electromagnetic force that can build up surprisingly high potentials across the device under test. Comparison of the drain currents and the temperatures of the corner and central parts cells of the transistor are shown in figure 5a and figure 5b. The temperature is almost evenly distributed and the currents through the transistors are equal during the first pulse. After some period of the power load, the temperature and breakdown voltage of the central transistor cell increase more significantly in comparison with the corner transistor cell because the corner cell is cooled more effectively. Therefore, the corner current becomes higher what is clearly seen at the beginning of the tenth UIS pulse. The inhomogeneous distributions of the temperature and current density are shown in figure 5c and figure 5d. Figure 5b shows comparison of the multipulse UIS for the MOSFET model using three-dimensional and one-dimensional thermal network. The inhomogeneous behavior along whole structure is neglected for one-dimensional simulation and the differences of the structure temperature and the drain breakdown voltage become more significant for the longer time simulation (tenth UIS pulse).

5. Conclusions

The super-junction vertical MOSFET was used for simulation of multipulse UIS test. The simulation approach helps to assess the device robustness by means of the evaluation of both temperature and current distributions in the MOSFET structures operating under critical conditions. The implemented three-dimensional thermal flow and distributed parameters of the MOSFET transistor provide more realistic simulation results. The advantages of the method are the relative simplicity of implementation, the speed of simulation time and the capability of full structure analysis.

6. References

[1] März M and Nance P 2000 Thermal modeling of power electronic systems *Application Note Infineon Technologies AG Munich*

[2] Fischer K and Shenai K 1996 Dynamics of power MOSFET switching under unclamped inductive loading conditions *IEEE Transactions on electron devices* **43** 1007

[3] Hefner H *et al* 1994 Component models for electrothermal network simulation *IEEE Transactions on Components Components, Packaging, and Manufacturing Technology* **17** 413

[4] Mawby P A *et al* 2001 New physics-based compact electro-thermal model of power diode dedicated to circuit simulation *IEEE Circuits and Systems* **3** 401

[5] Laprade A *et al* 2003 A new PSPICE electro-thermal subcircuit for power MOSFETs, *Application Note 7532 Fairchild*

[6] Bagnoli P E 1998 Thermal resistance analysis by induced transient (TRAIT) method for power electronic devices thermal characterization—part I fundamentals and theory *IEEE Transactions on Power Electronics* **13** 1220

[7] Jancke R *et al* 2010 Simulation of electro-thermal interaction, *Electronics System Integration Technology Conf., ESTC* 1

[8] Fujihira T 1997 Theory of semiconductor superjunction devices *Appl. Phys.* **36** 6254

[9] Moens P *et al* 2011 UltiMOS: a local charge-balanced trench-based 600V super-junction device *Int. Symp. on Power Semiconductor Devices & ICs, ISPSD* 304

[10] Rhayem J *et al* 2012 Novel 3D electrothermal robustness optimization approach of super junction power MOSFETs under unclamped inductive switching *IEEE SEMI-THERM Symp.* 69

[11] Verilog-A language reference manual analog extensions to Verilog HDL 1996

[12] Chvala A *et al* 2012 Power transistor models with temperature dependent parasitic effects for SPICE-like circuit simulation *IEEE Int. Conf. on Microelectronics* 255

Acknowledgments

This work has been done with support of 7FP project SMAC, no. 288827 and grant VEGA 1/0866/11.

Journal of Physics: Conference Series **494** (2014) 012007 doi:10.1088/1742-6596/494/1/012007

Modelling Solar Cells with Thermal Phenomena Taken into Account

K Górecki, P Górecki and K Paduch
Gdynia Maritime University, Department of Marine Electronics, Morska 83, Gdynia, Poland

E-mail: gorecki@am.gdynia.pl

Abstract. The paper is devoted to modelling properties of solar cells. The authors' electrothermal model of such cells is described. This model takes into account the influence of temperature on its characteristics. Some results of calculations and measurements of selected solar cells are presented and discussed. The good agreement between the results of calculations and measurements was obtained, which proves the correctness of the elaborated model.

1. Introduction

In recent years, more and more attention is paid to renewable energy sources. To this class of energy sources belong e.g. photovoltaic systems [1 - 4]. The main components of these systems are solar cells, in which radiation energy (for example solar radiation) is converted into electrical energy. The solar cell comprises a p-n junction, whose properties strongly depend on temperature [5, 6].

In the design and analysis of electronic circuits the computer software is commonly used. Nowadays the most popular is program SPICE [7, 8]. For example, the book [9] is dedicated to the problem of modelling photovoltaic systems using SPICE. This book presents a method of modelling particular components of a photovoltaic system - from solar cells to inverters and batteries. Unfortunately, the models presented in the cited book do not take into account many important physical phenomena, such as self-heating, or they describe some phenomena in a highly simplified manner, e.g. the influence of changes in the ambient temperature on the characteristics of a solar cell.

In turn, in the paper [8] the model of the photovoltaic module dedicated for the SPICE software is proposed. The equations describing characteristics of the module take into account the influence of temperature on its characteristics. In this model the temperature is a parameter and it does not change during the analysis, therefore it can be interpreted as the ambient temperature. In the considered model the dependence of the photocurrent on the power density of the radiation is also skipped.

As it can be seen in papers [1, 2, 9, 10], the solar cell temperature during its operation can be significantly higher than the ambient temperature due to self-heating or heating by the light source [11 - 13]. The increasing temperature of the solar cell could shorten its lifetime and reduce the value of the voltage at the terminals of solar cells [9].

In the paper [14] the electrothermal model of the solar cell for the SPICE software is proposed. This model takes into account self-heating phenomena, but it skips an increase of temperature of this element resulting from the exchange of a part of the energy of radiation on the solar cell area into heat.

In this paper, which is an extended version of the paper [14], an electrothermal model of solar cell for SPICE is proposed. This model takes into account the influence of the photovoltaic effect and drawbacks of thermal phenomena on the characteristics of this device. The presented model is elaborated and verified for silicon monocrystalline solar cells.

In the second section, the elaborated model is described, whereas in the third section the selected results of measurements and calculations are presented.

2. Model description

The worked out by the authors electrothermal model of the solar cell takes into account photovoltaic phenomena, self-heating and a temperature increase resulting from the thermal radiation falling on the surface of this solar cell. The network representation of the elaborated model is shown in figure 1.

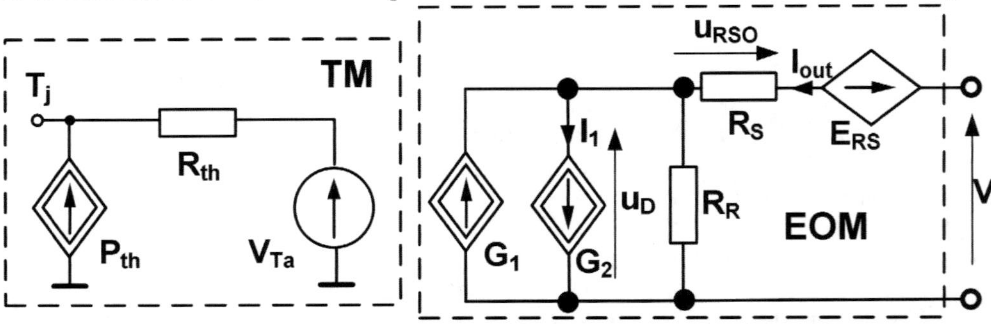

Figure 1. Network representation of the electrothermal model of a solar cell.

The elaborated model consists of two parts: the electro-optical model (EOM) and the thermal model (TM). In the EOM, the controlled current source G_1 represents the photo current described by the following formula

$$I_{FO} = P \cdot S \cdot \eta \cdot x \cdot \left[1 + \alpha_T \cdot (T_j - T_0)\right] \tag{1}$$

where P denotes the power density of the lighting radiation, S - active area of the solar cell, η – efficiency of photovoltaic conversion, x - the coefficient equal to 1 V^{-1}, α_T – temperature coefficient of changes of the photo current, T_j – internal temperature of the solar cell, T_0 - reference temperature.

In turn, the controlled current source G_2 represents the current of the p-n junction described by the classical Shockley's model of the form

$$I_1 = J_o \left(\frac{T_j}{T_0}\right)^2 \cdot \exp\left(-\frac{U_{go}}{h \cdot T_j}\right) \cdot S \cdot \left[\exp\left(\frac{u_D}{n \cdot h \cdot T_j}\right) - 1\right] \tag{2}$$

where J_0 denotes the parameter of the saturation current density, n – emission factor of the p-n junction, U_{go} – voltage corresponding to the band-gap of silicon, u_D – junction voltage, and h is the quotient of the Boltzmann constant through the electron charge.

The resistor R_R models leakage of the junction, while R_S means series resistance of the solar cell in temperature T_0. The controlled voltage source E_{RS} models the influence of temperature on series resistance of the solar cell by linear dependences of the form

$$E_{RS} = U_{RSO} \cdot \alpha_{RS} \cdot (T_j - T_0) \tag{3}$$

where α_{RS} denotes the temperature coefficient of changes of series resistance, whereas U_{RSO} – the voltage on the resistor R_S.

The value of the internal temperature of the solar cell is calculated in the thermal model (TM). In this model, the current source P_{th} represents power dissipated on the series resistance, on the leakage resistance and the power of radiation absorbed by the solar cell. The current of the controlled current source P_{th} is described by

$$P_{th} = P \cdot a \cdot S_1 + (V_{out} - u_D) \cdot I_{out} + u_D \cdot I_1 \tag{4}$$

where a denotes the ratio of conversion of the radiation energy into heat, S_1 – area of the solar cell, while V_{out} and I_{out} are the output voltage and the output current of the solar cell, respectively. R_{th} models the thermal resistance of solar cells, whereas V_{Ta} represents the ambient temperature.

3. Estimation of the model parameters

The model presented in section 2 is described by means of 11 parameters. The values of these parameters are indispensable to calculate the solar cell characteristics by means of this model. In this section the manner of estimation of these parameters values is proposed. In this manner the idea of the local estimation, described among other things in the paper [15] is used.

The value of the reference temperature T_0 is chosen by the user and means the temperature in which the values of the remaining parameters are estimated. The area of the active part of the solar cell S and the entire area of this solar cell S_1 are estimated on the basis of the geometrical dimensions of the suitable areas of the solar cell and formulas that are well-known from geometry.

The efficiency of the conversion η is calculated on the basis of the measurement of the short circuit current of the solar cell I_{SC} at the well-known value of the power density of radiation P_R. The value η is given by the formula

$$\eta = \frac{I_{SC}}{S \cdot P_R \cdot x} \tag{5}$$

The temperature coefficient of changes of the photocurrent α_T can be estimated after measuring the values of the short-circuit current I_{SC} and I_{SC1} corresponding to the temperature values T_0 and T_1, respectively at the identical value of the power density of radiation P_R. The value of the coefficient α_T can be calculated using the following equation

$$\alpha_T = \frac{I_{SC} - I_{SC1}}{(T_0 - T_1) \cdot I_{SC}} \tag{6}$$

In order to estimate the values of parameters J_0, n, R_S, α_{RS} and R_R the measurement of the characteristic i(v) of the not lighted up solar cell in temperatures T_0 and T_1 is indispensable. The slope of this characteristic in the range of small voltages (typically smaller than 0.15 V) is equal to R_R resistance. On the basis of the coordinates of points $A(v_1, i_1)$ and $B(v_2, i_2)$ laying on the linear section of the current-voltage characteristic i(v) presented in the linear-logarithmic scale one can calculate the value of the coefficient n from the formula

$$n = \frac{v_1 - v_2}{h \cdot T_0 \cdot \ln(i_1 / i_2)} \tag{7}$$

and the value of the parameter J_0 using the formula

$$J_0 = \frac{i_1}{S} \cdot \exp\left(-\frac{v_1}{n \cdot h \cdot T_0}\right) \cdot \exp\left(\frac{U_{go}}{h \cdot T_0}\right) \tag{8}$$

The series resistance of the solar cell R_S is estimated on the basis of the measurement of the coordinates of the point $C(v_3, i_3)$ laying in the range of high currents of the solar cell. In the calculations the equation of the following form is used

$$R_S = \frac{v_3 - n \cdot h \cdot T_0 \cdot \ln\left(i_3 \Big/ \left(J_0 \cdot S \cdot \exp\left(-\frac{U_{go}}{h \cdot T_0}\right)\right)\right)}{i_3} \tag{9}$$

In order to estimate the value of the coefficient α_{RS} one ought to estimate the values of the resistance R_S for two values of temperature T_0 and T_1, and then to use the formula

$$\alpha_{RS} = \frac{R_S(T_1) - R_S(T_0)}{(T_1 - T_0) \cdot R_S(T_0)} \tag{10}$$

The value of the thermal resistance of the solar cell R_{th} is estimated with the use of the method described in the paper [16], using the coordinates of points laying on the static characteristic of the not illuminated forward biased solar cell. The value of the parameter a is estimated at the lighting of the solar cell from the light source of the well-known power density P and at the small value of the forward current of the solar cell. On the basis of the measurement of the voltage on the solar cell once the light source is switched on at the steady-state, the increase of the internal temperature of the solar cell ΔT can be estimated. The value of the parameter a is calculated from the formula

$$a = \frac{\Delta T}{R_{th} \cdot P \cdot S_1} \tag{11}$$

4. Investigation results

In order to verify the correctness of the elaborated model, the calculations and measurements of the selected characteristics of the solar cells operating at different temperatures and with different values of radiation intensity are performed. The measurements were made for monocrystalline silicon solar cells of different sizes. The thickness of the solar cells with the PCB, on which they were located, was equal to 1.5 mm. During the investigations these solar cells were located in the light-tight thermostat of the volume equal to about 0.25 m^3. The incandescent bulbs were the radiation source. The spectrum of the emitted light is close to the spectrum of the solar light [17]. Both the isothermal (without self-heating) and non-isothermal (with self-heating) characteristics of the solar cells were calculated and measured.

In order to eliminate the influence of self-heating on the result of the measurements of isothermal characteristics, the impulse method of measurements was used. The bulbs were switched on in time equal to about 5 s and then the values of the current and the voltage on the output terminals of the solar cell were measured. Between the following measurements, at least a five-minute-pause appeared to assure cooling of the solar cell to the ambient temperature. The value of the current was regulated by means of the decadal resistor, which was the load of the solar cell.

The results of isothermal calculations and measurements are presented in figures 2-5. In all figures presented in this section points mark the results of measurements, dashed lines – the results of calculations obtained with the model from the paper [9], while solid lines – the results of calculations obtained with the authors' model.

The values of electrical and optical parameters of the solar cell Conrad 0.5V/400mA estimated with the method described in the section 3 are collected in the Table 1.

Table 1. Value of the authors' model parameters of the solar cell Conrad 0.5V/400mA.

parameter	S [m^2]	S_1 [m^2]	η	T_0 [K]	α_T [1/K]	J_0 [A/m^2]
value	12×10^{-4}	35×10^{-4}	0.115	300	1.9×10^{-3}	4.17×10^{12}
parameter	n	R_S [Ω]	α_{RS} [1/K]	R_R [kΩ]	a	
value	1	0.3	3×10^{-3}	10	0.8	

For example, figure 2 shows the current-voltage characteristics of the solar cell Conrad 0.5V/400mA for the selected values of the ambient temperature. The presented results were obtained at illumination of the investigated solar cells with one bulb, which corresponds to the power density of radiation on the surface of the solar cell equal to 416 W/m^2.

As it can be seen, the output voltage of the solar cell is a decreasing function of the output current and temperature. Together with the temperature rise, the value of the short circuit current increases. Comparing the obtained results of calculations and measurements, it is visible that the good agreement of the results of calculations with the electrothermal model and measurements, ensuring the accuracy of the elaborated model, was obtained. From the calculations performed with the use of the literature model, the understated values of the output voltage were obtained in the range of low currents and high temperature.

In turn, in figure 3 the characteristics of the same solar cell obtained at illumination with two bulbs are presented. In this case the power density of radiation of the surface of the solar cell amounts to 772 W/m^2.

As it can be noticed, an increase in the illuminance of the solar cell caused an increase in the value of its short circuit current. In this figure, it is also visible that the manner of modelling the influence of temperature on the characteristics of the solar cell used in the literature model is incorrect. On the other hand, the electrothermal model assures the good agreement between the calculated and measured characteristics in all the considered range of changes of temperature and the output current.

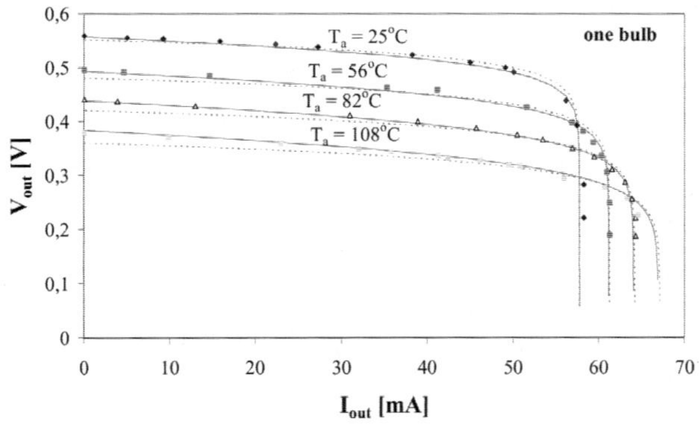

Figure 2. Calculated and measured current-voltage characteristics of the solar cell Conrad 0.5V/400mA in the case when one bulb lights.

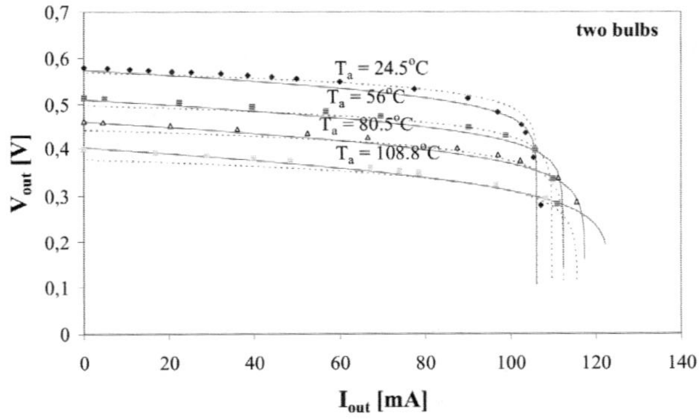

Figure 3. Calculated and measured current-voltage characteristics of the solar cell Conrad 0.5V/400mA in the case when two bulbs light.

In figure 4 – the dependences of the output power on the load current for the considered solar cell are presented.

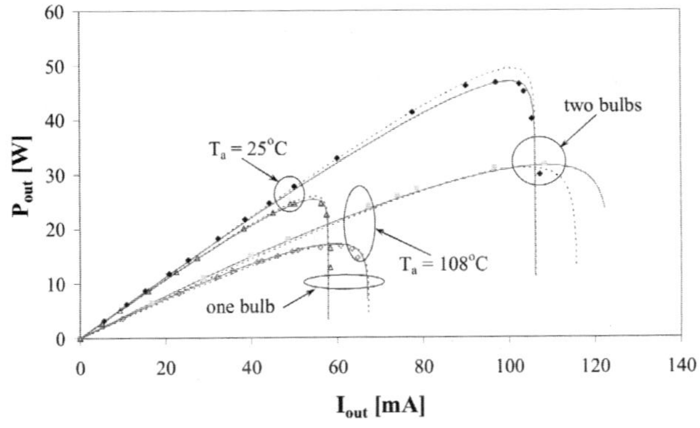

Figure 4. Calculated and measured dependences of the output power on the load current for the solar cell Conrad 0.5V/400mA.

Journal of Physics: Conference Series **494** (2014) 012007 doi:10.1088/1742-6596/494/1/012007

As it should be anticipated, while lighting the solar cell with two bulbs, almost twice higher maximum values of the output power were obtained. The temperature increase from 25 to 108 °C causes a fall in the value of this power by even about 40% and an increase of the maximum value of the output current even by over 10%.

In figure 5 the characteristics of two monocrystalline solar cells of the different surface obtained from the measurements and calculations performed with the use of the electrothermal model are compared. The solar cell Conrad 0.5V/400mA has dimensions 76x46 mm, whereas the solar cell 0.5V/100mA - 46x26 mm. As it can be noticed, for both the considered solar cells the good agreement between the calculated and measured characteristics was obtained in a wide range of temperature changes. However, the achievement of this agreement demanded taking into account the fact, that the quotient of the areas S of both the investigated solar cells was not equal to the quotient of the total areas of these elements, but it was equal to the quotient of the nominal values of the output current of the considered solar cells. The observed relation results from the fact that only some part of the surface of the investigated solar cells is the active area, and the banks of the element do not participate in generation of the photocurrent.

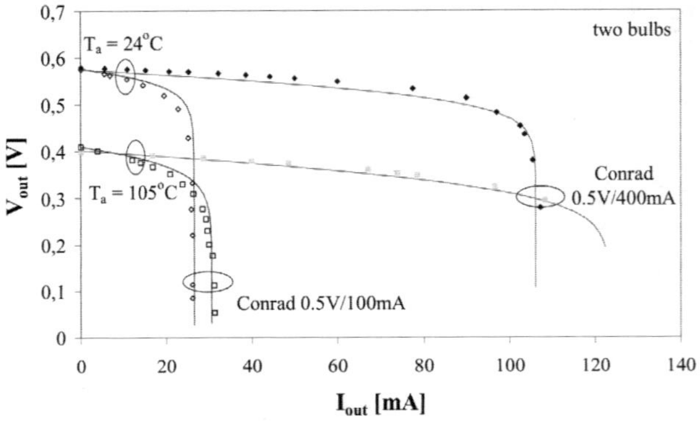

Figure 5. Calculated and measured current-voltage characteristics of the solar cells Conrad 0.5V/400mA and Conrad 0.5V/100mA.

In figure 6 the calculated and measured characteristics of the solar cell Conrad 0.5V/400 obtained for the not lighted up device at different values of the ambient temperature are presented. As one can notice, together with the temperature rise the shift of the considered characteristics in the left direction is observed, similarly as it is observed for p-n or m-s junctions [5]. The temperature coefficient of changes of the forward voltage equals -1.9 mV/K.

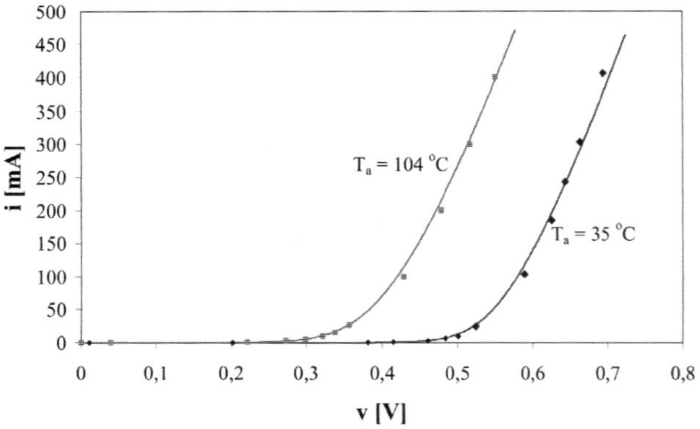

Figure 6. Calculated and measured dark current-voltage characteristics of the solar cells Conrad 0.5V/400mA.

In figure 7 the influence of the cooling conditions of the solar cell on its current-voltage characteristics is illustrated. In this figure the curve a denotes the characteristic corresponding to ideal cooling of the investigated device (the isothermal measurement), the curve b - the characteristic obtained in the steady-state for the solar cell situated in the closed light-tight chamber of the volume equal approx. to 0.25 m^3, whereas the curve c - the characteristic obtained in the steady-state for the solar cell situated in the open chamber. In all the cases the solar cell is lighted up by two bulbs of the nominal power 75 W, which corresponds to the power density of radiation equal to 772 W/m^2. In the calculations, the results of which are shown in figure 7, the following values of thermal resistance of the solar cell are used: $R_{th} = 0$ for the isothermal characteristics, $R_{th} = 40.3$ K/W for the solar cell situated in the closed chamber and $R_{th} = 14.3$ K/W for the solar cell situated in the open chamber.

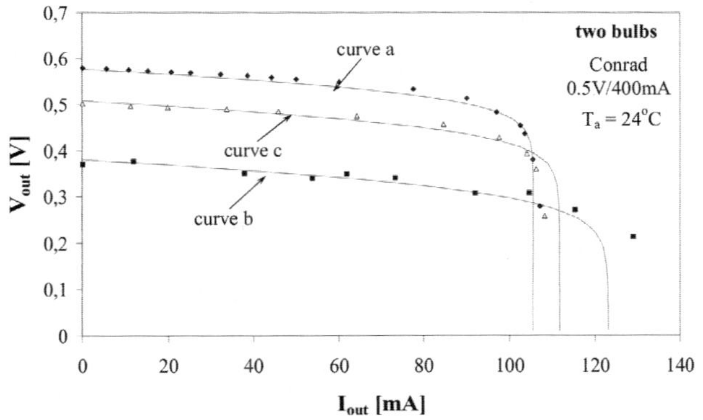

Figure 7. Calculated and measured current-voltage characteristics of the solar cells Conrad 0.5V/400 mA obtained at different cooling conditions.

As can be observed, worsening of the cooling conditions (the increase in value R_{th}) results in a decrease in the value of the voltage on the output of the solar cell and with an increase in the value of the short circuit current. It is good to notice that as a result of self-heating and absorption of the heat generated by the light source, the change in the course of characteristics of the solar cell has the identical character as changes in the course of these characteristics due to the change of the value of the ambient temperature. The values of the internal temperature of the solar cell obtained from the calculations for the device situated in the open chamber amount to about 55°C, and for the device situated in the closed chamber – up to 111°C.

5. Conclusions

In the paper the form of the electrothermal model of the solar cell is proposed. This model is dedicated to be implemented in the form of the subcircuit for the SPICE software. It takes into account self-heating phenomena and heating resulting from the exchange of energy of the part of the absorbed radiation into heat and changes of the ambient temperature. The presented manner of estimation of the value of model parameters is very simple to realize.

The presented in the previous section initial results of the experimental verification of the worked out model confirm its usefulness in modelling characteristics of solar cells over a wide range of changes in the output current of the solar cell, the ambient temperature, cooling conditions and active areas of the solar cell. The presented model assures the better agreement between the results of calculations and measurements than the classical literature model.

6. References

[1] Rashid MH 2007 Power Electronic Handbook *Academic Press, Elsevier*
[2] Klugmann-Radziemska E 2010 Fotowoltaika w teorii i praktyce *Wydawnictwo BTC, Legionowo*
[3] Ersam T, Kimball J, Krein P, Chapman P and Midya P 2006 Dynamic Maximum Power Point Tracking of Photovoltaic Arrays Using Ripple Correlation Control *IEEE Transaction on Power*

Electronics **21** (5) 1282-1291

[4] Villalva MG, Gazoli JR and Filbo ER 2009 Comprehensive Approach to Modeling and Simulation of Photovoltaic Arrays *IEEE Transactions on Power Electronics* **24** (5) 1198-1208

[5] Janke W 1992 Zjawiska termiczne w elementach i układach półprzewodnikowych *WNT, Warszawa*

[6] Singh J 2000 Semiconductor devices: basic principles *John Wiley&Sons*

[7] Rashid M and Rashid H 2006 SPICE for Power Electronics and Electric Power *Taylor and Francis Group, LLC*

[8] Gadjeva E and Hristov M 2013 Behavioral Parametrized SPICE Models of Photovoltaic Modules *Proceedings of the 20th International Conference Mixed Design of Integrated Circuits and systems MIXDES* 355-359

[9] Castaner L and Silvestre S 2002 Modelling photovoltaic systems using Pspice *John Wiley&Sons*

[10] Znajdek K, Walczak S and Sibiński M 2011 Analiza parametrów modułów PV różnych typów w warunkach rzeczywistych *X Krajowa Konferencja Elektroniki KKE Darłówko Wschodnie* 128-133

[11] Garcia Alonso MC and Balenzategui JL 1997 Estimation of photovoltaic module yearly temperature and performance based on Nominal Operation Cell Temperature calculations *Renevable Energy* **29** 1997-2010

[12] Lee WM, Infield DG and Gottschalg R 2001 Thermal modelling of building integrated PV systems *17th European Photovoltaic Solar Energy Conference and Exhibition Munchen* 1447.

[13] Marańda W and Piotrowicz M 2010 Extraction of thermal model parameters for field-installed photovoltaic module *27th International Conference on Microelectronics Proceedings (MIEL)* 153 – 156

[14] Górecki K, Górecki P and Paduch K 2013 Modelling of solar cells characteristics with thermal phenomena taken into account *Proceedings of Microtherm 2013 Microtechnology and Thermal Problems in Electronics Łódź* 298-303

[15] Zarębski J and Górecki K 2002 Parameters Estimation of the D.C. Electrothermal Model of the Bipolar Transistor *International Journal of Numerical Modelling Electronic Networks, Devices and Fields* **15** (2) 181-194

[16] Zarębski J and Górecki K 2008 A Method of the Thermal Resistance Measurements of Semiconductor Devices with P-N Junction *Measurement* **41** (3) 259-265

[17] Cheng YK and Cheng KWE 2006 General Study for using LED to replace traditional lighting devices *2nd International Conference on Power Electronics Systems and Applications ICPESA '06 Hong Kong* 173 - 177

Journal of Physics: Conference Series **494** (2014) 012008 doi:10.1088/1742-6596/494/1/012008

A Semiconductor Device Thermal Model Taking into Account Non-linearity and Multhipathing of the Cooling System

K Górecki and J Zarębski

Gdynia Maritime University, Department of Marine Electronics, Morska 83, Gdynia, Poland

e-mail: gorecki@am.gdynia.pl

Abstract. The paper is devoted to modelling thermal properties of semiconductor devices at the steady state. The dc thermal model of a semiconductor device taking into account the multipath heat flow is proposed. Some results of calculations and measurements of thermal resistance of a power MOSFET operating at different cooling conditions are presented. The obtained results of calculations fit the results of measurements, which proves the correctness of the proposed model.

1. Introduction

Ensuring effective cooling of electronic components is one of the most important challenges the designers and constructors of electronic and power electronics devices are facing. The designer usually has information about the estimated value of power dissipated in discrete electronic components and the imposed restrictions concern, e.g. a permissible value of input power from the power source and the maximum size and shape of the system casing. Under these conditions, the cooling system components are often intuitively chosen. The selection of components of the heat flow path is complicated and difficult due to multipath heat transport from the die to the surrounding [1, 2]. In order to determine thermal properties of the electronic components, their thermal models are formulated. The structure of such a model, taking into account the multipathing heat transport is shown in figure 1 [2].

As you can see, the heat can flow in various ways, using three mechanisms of its transfer: conduction, convection and radiation. Since the description of the phenomena associated with the heat flow is very complex, the practical verification of the design of the cooling system is made experimentally by constructing a prototype. This often results in long time designing or oversizing significantly cooling system components, leading to an unjustified increase in the cost of the device.

The commonly used thermal models have a form of the electric circuit based on the electric-thermal analogy [3 - 6]. Most frequently the elements of the heat transfer path, indicated in figure 1, are represented by the RC network of the Cauer or Foster structure [3 - 6]. Due to the dependence of the parameters describing the heat flow through each element of the path of the heat flow on temperature, the thermal model is generally non-linear [3]. The values of resistances and capacitances existing in the thermal model are typically obtained from the measurements realised with the use of optical or electrical methods [7- 11]. Many papers, for example [8 - 14], present the results of measurements of the thermal resistance of semiconductor devices, in which the influence of the power dissipated in the investigated device and the construction of the cooling system of this device on its thermal resistance is clearly visible.

In the literature, such as [15 - 19] only the selected aspects of modelling the thermal behaviour of electronic components are considered. Moreover, empirical formulas describing dependencies of the device transient thermal impedance on the dissipated power for one chosen cooling condition of the device are described [3].

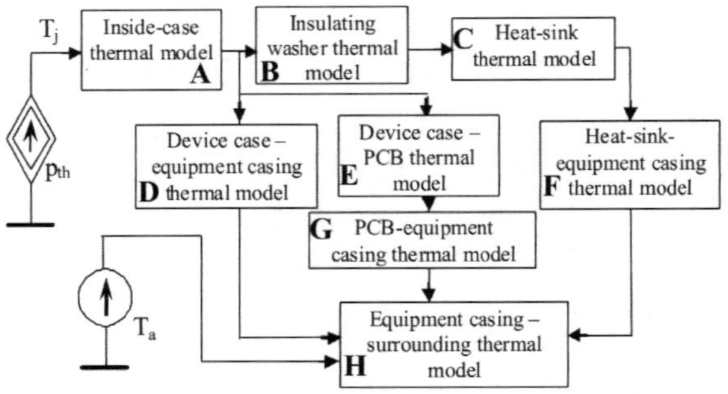

Figure 1. General form of the thermal model of a semiconductor device constituting a component of an electronic device mounted in the casing.

In this paper, which is an extended version of the paper [13], thermal properties of the circuit consisting of the power transistor mounted on the aluminium heat-sink with insulating washer and the Peltier module are analysed. Additionally, the thermal properties of this transistor operating without any heat-sink or located on the heat-sink without the Peltier module are considered.

In section 2 the measurement setup used to measure the thermal parameters of semiconductor devices is presented. The form of the thermal model is described in section 3. Section 4 presents the results of investigations of the power MOS transistor operating at different cooling conditions.

2. Measurement setup

The results of measurements presented in this paper are obtained by means of the electric method of measurements of transient thermal impedance using the cooling curve [11, 20]. The voltage on the forward biased body diode is used as a thermally-sensitive parameter. The measurements were realised in the authors' measurement setup, described in the paper [12]. The block diagram of this measurement setup is shown in figure 2.

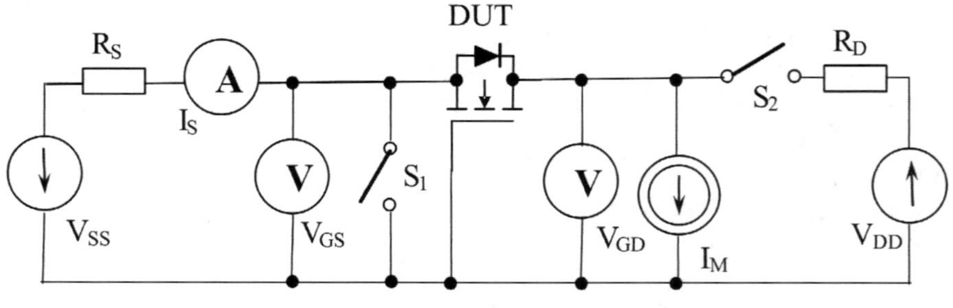

Figure 2. Block diagram of the measurement setup.

In the presented measurement setup, the examined transistor (DUT) operates in the common gate configuration. The source current is regulated by means of the voltage source V_{SS} and the resistor R_S, and its value I_S is measured by the ammeter. The value of the gate-drain voltage V_{GD} is regulated by the voltage source V_{DD}. The values of voltages V_{GS} and V_{GD} are measured by means of voltmeters. The measurement of thermal resistance is realised in 3 stages. First, the thermometric characteristic $V_{GD}(T)$ of the body diode at the forward current equal to I_M is measured. During the calibration the switch S_2 is off and the switch S_1 – is on. The value of the current I_M is equal to 10 mA. At this value of I_M the thermometric characteristic is linear [12]. In the second stage, the examined transistor self heats as a result of dissipation of the power $P_H = I_S \cdot (V_{GS} - V_{GD})$ in it. This stage lasts until the steady state is obtained and the voltage V_{GS} practically does not change. In the third stage the transients of the voltage V_{GD} on the body diode are registered from the moment t = 0, in which the switch S_1 is turned-on and the switch S_2 is turned-off, until the steady state is obtained. The PC registers values of the voltage on the body diode and controls the switches S_1 and S_2.

On the basis of the measured time course $V_{GD}(t)$, the thermometric characteristic $V_{GD}(T)$ and the value of the power P_H at the end of the second stage of the measurement, the course of the transient thermal impedance $Z_{th}(t)$ is calculated using the following formula:

$$Z_{th}(t) = \frac{V_{GD}(t) - V_{GD}(T_a)}{P_H \cdot F} \tag{1}$$

where T_a denotes the ambient temperature and F - the slope of the thermometric characteristic $V_{GD}(T)$. The thermal resistance R_{th} is equal to the value of $Z_{th}(t)$ at the steady state.

Of course, it is possible to measure the value of the thermal resistance using any dc method, but the range of currents and voltages, in which this method could be used is strongly limited [8, 14].

3. Form of the thermal model

In the paper the cooling system of the power MOS transistor containing a heat-sink and a Peltier module is considered. The formulated dc thermal model of the considered system of the general form is shown in figure 3. In this model, the controlled current source G_{th} represents power dissipated in the modelled device, the voltage source V_{Ta} - ambient temperature, whereas non-linear resistors – thermal resistance of the components of the heat flow path.

On the basis of the dependencies specified by the producers describing thermal resistances of individual components of the heat flow path and the measurements made by the authors, the analytic expressions describing the dependence of the thermal resistance of individual components of the heat flow path are formulated.

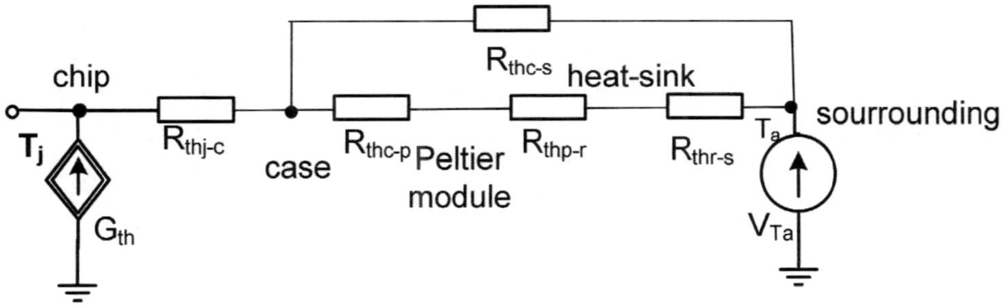

Figure 3. DC thermal model of a semiconductor device mounted on a heat-sink.

The thermal resistance R_{thj-c} is constant and its value is equal to the value given by the producer in the catalogue card. The values of the thermal resistance between the case and the surrounding R_{thc-s} is estimated from the measurements of R_{thj-a} of the transistor operating without any heat-sink as the difference between R_{thj-a} and R_{thj-c}.

In turn, the thermal resistance of the isolating washer R_{thc-a} is usually very small and with the correctly performed connection does not exceed typically 1 K/W. Therefore, in the thermal model one can arbitrarily accept the top value of this parameter R_{thc-a} = 1 K/W. The thermal resistance of the isolating washer can be given on the basis of the values of the thermal conductance and geometrical dimensions of the element case given by the producer or based on the measurements of the thermal resistance R_{thj-a} of the transistor mounted on the heat-sink for two cases: R_{th1} - without the isolating washer and R_{th2} - with such a washer. The difference R_{th2} - R_{th1} is equal to the value of the thermal resistance R_{thc-a}.

As it was shown e.g. in [21, 22] the thermal resistance of the power transistor can change in an essential manner e.g. in the function of the pressure exerted on the washer isolating the heat-sink and the device case. The value R_{thj-a} of the examined device in the considered range of changes of pressure p changes even twice, but for the value of the pressure p higher than 1.2 N/mm^2, which corresponds to tight screwing of the screw, the thermal resistance is practically constant. Then, for the correctly performed connection of the device to the heat-sink the value of R_{thc-a} is practically constant.

In turn, the thermal resistance between the heat-sink and the surrounding R_{thr-s} can be obtained from the catalogue dependences of the thermal resistance of the heat-sink on its length. In the case, when such dependence is not accessible, the value R_{thr-s} can be obtained on the basis of the measurements of the thermal resistance R_{thj-a} of the transistor installed on the heat-sink. For this situation, typically the value of R_{thc-s} is omissible high. In such a case the value R_{thr-s} can be calculated from the dependence:

$$R_{thr-s} = R_{thj-a} - R_{thj-c} - R_{thc-p} - R_{thp-r} \tag{2}$$

It should be taken into account that as it results e.g. from [12], the value of R_{thr-s} depends on the spatial orientation of the heat-sink and the values of this parameter at different orientations of the heat-sink can differ from one another even by about 20%.

The value of the thermal resistance between the Peltier module and the heat-sink R_{thp-r} can be obtained from the measurements of R_{thj-a} of the transistor situated on the heat-sink for two cases: with the Peltier module R_{th3} and without the Peltier module R_{th4}. Then

$$R_{thp-r} = R_{th3} - R_{th4} \tag{3}$$

The value R_{thp-r} is a decreasing function of both the power p_{th} dissipated in the examined transistor and the power p_p supplying the module. Of course, on the basis of the measurements of the mentioned dependences it is possible to describe thermal properties of the selected Peltier module.

4. Results

In order to verify the correctness of the proposed thermal model, the measurements and calculations of the thermal resistance of the power MOS transistor IRF530 operating at different cooling conditions are performed. The considered thermal model has the form shown in figure 3. On the basis of the measurement performed by the authors, the following values of model parameters were estimated: the thermal resistance between the inside and the case in compliance with the catalogue card [23] is R_{thj-c} = 1.9 K/W. The thermal resistance of the connection between the case and the heat-sink filled by silicone is equal to 0.5 K/W.

Several cooling system configurations are explored. In the simplest case, when the transistor operates without the heat-sink, its thermal resistance is a function of the dissipated power [3, 12, 13]. For the vertically situated transistor in the case TO-220, the dependence $R_{thc-a}(p_{th})$ can be described with the following empirical formula

$$R_{thrc-a} = R_4 + R_5 \cdot \left(1 - \exp\left(-p_{th0}/d\right)\right) \tag{4}$$

For the considered cooling system the values of parameters in equation (4) are: R_4 = 44.8 K/W, R_5 = 8 K/W, d = 2.5 W.

For example, for the transistor situated on the heat-sink without the additional isolating washer the thermal resistance between the heat-sink and the surrounding R_{thr-s} is non-linear, i.e. its value depends on the power p_{th} dissipated in the transistor according to the formula

$$R_{thr-s} = R_0 + R_1 \cdot \exp\left(-\left(p_{th} - p_0\right)/a\right) \tag{5}$$

For the considered cooling system R_0 = 1.9 K/W, R_1 = 1.5 K/W, p_0 = 12 W, a = 8 W.

The operation of the transistor IRF530 separated from the same heat-sink through the Peltier module is also considered. The thermal resistance of the connection between the case of the device and the module is 0.5 K/W. In turn, the thermal resistance of the Peltier module depends both on the temperature difference between its both sides, and hence on the power p_{th} dissipated in the examined device, and on the power p_p supplying the Peltier module. The thermal resistance R_{thp-r} is described by the formula

$$R_{thp-r} = R_2 + R_3 \cdot \exp\left(-p_p/b\right) + R_4 \cdot \exp\left(-\left(p_{th} - p_1\right)/c\right) \tag{6}$$

For the considered transistor, the values of the parameters in the equation (6) are as follows: $R_2 = 4.7$ K/W, $R_3 = 1.8$ K/W, $R_4 = 1$ K/W, $p_1 = 5$ W, b = 4 W, c = 3.2 W.

In figures 4 - 7 the calculated and measured dependences of the thermal resistance between the interior of the investigated transistor and the surrounding on the dissipated power (figures 4-6) or on the power supplied to the Peltier module (figure 7) at different cooling conditions of the considered transistor, are presented. In these figures points mark the results of measurements, while lines – the results of calculations obtained on the basis of the model proposed in this paper.

Figure 4 refers to the transistor operating without any heat-sink. In this case a dominant component of the thermal resistance of the transistor is the thermal resistance between the case and the surrounding R_{thc-a}. The increasing character of the dependence $R_{thc-a}(p_{th})$, which testifies the dominant role of the thermal conductivity in dissipation of the heat, determines the shape of the obtained curve. As it can be noticed, the range of changes of the value of the thermal resistance of the transistor is in the considered case not high and does not exceed 10%.

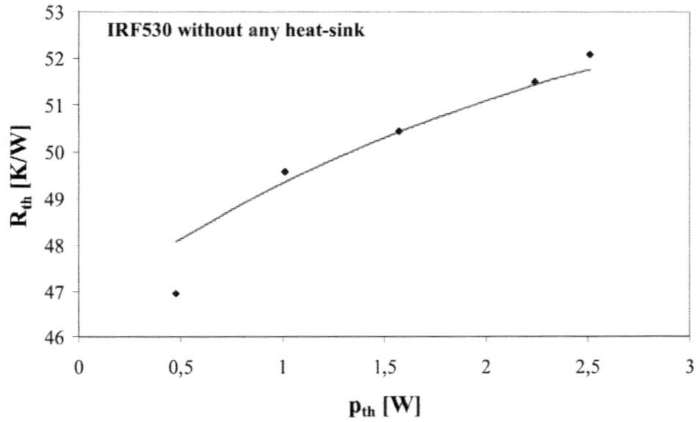

Figure 4. Calculated and measured dependences of thermal resistance of the IRF530 transistor operating without any heat-sink on the dissipated power.

In turn, figure 5 refers the transistor situated directly on the heat-sink. In comparison to the values obtained for the transistor without any heat-sink the almost tenfold fall in value of the thermal resistance of the investigated transistor is obtained. In the considered case, the thermal resistance between the case and the surrounding R_{thc-a} is omissible large in relation to the thermal resistance between the heat-sink and the surrounding R_{thr-a}. The decreasing character of the dependence $R_{th}(p_{th})$ shows that convection is a dominant mechanism of heat abstraction from the transistor.

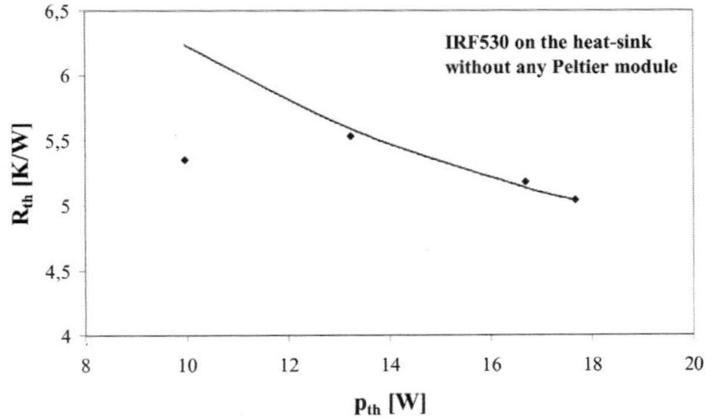

Figure 5. Calculated and measured dependences of thermal resistance of the IRF530 transistor situated on the heat-sink on the dissipated power.

In figure 4 the calculated and measured dependences of thermal resistance of the examined transistor on the power dissipated in the transistor at the zero-value of the power supplying the Peltier module are compared, while in figure 5 the dependences of thermal resistance of this transistor on the power p_p are shown.

In both the considered cases the visible fall in the value of thermal resistance at an increase of the power dissipated in the transistor and the power supplying the Peltier module is obtained. It is worth noticing that the use of the not supplied Peltier module caused an increase over twice higher in the value of thermal resistance. It is the result of the introduction of the additional component of thermal resistance between the case of the device and the heat-sink. The forced cooling of the transistor by the use of the Peltier module allows decreasing the thermal resistance of the transistor even by about 20%, but it remain constant on the considerably higher level than in the case, when the Peltier module is not used.

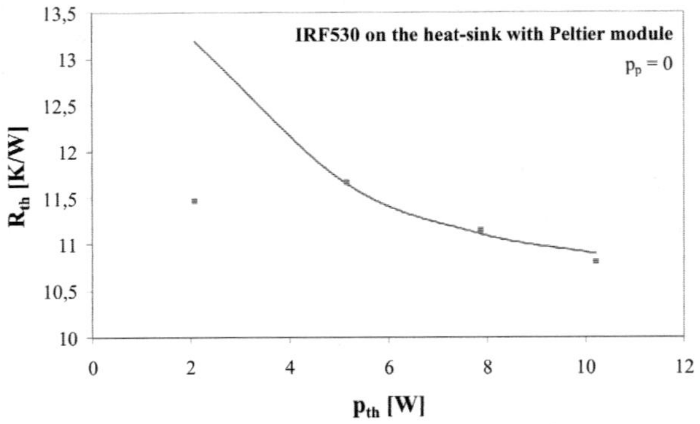

Figure 6. Measured and calculated dependence of thermal resistance of the transistor on the dissipated power.

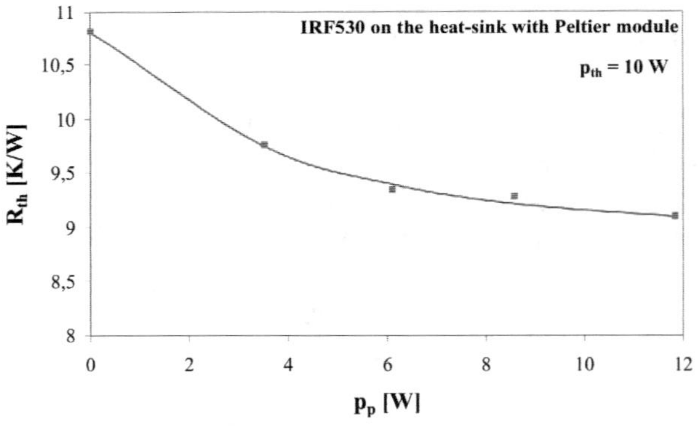

Figure 7. Measured and calculated dependence of thermal resistance of the transistor on the power supplying the Peltier module.

As it can be noticed, a satisfied agreement between the measured and calculated dependences $R_{th}(p_{th})$ and $R_{th}(p_p)$ is obtained in the range of high value of dissipated power. It testifies to the correctness of the proposed thermal model and the correctness of the estimation of the parameters values of this model.

5. Conclusions

In the paper the thermal model of power semiconductor devices taking into account the multipath and nonlinearity of the phenomena describing the heat transport is proposed. The detailed considerations are presented on the example of the power MOS transistor IRF530. The results of calculations and measurements shown in the paper prove the correctness of the proposed model. The parameters of this model are estimated on the basis of the catalogue data or from the simple measurement, which can be performed by the user. The obtained results of the verification of the model both at free and forced cooling testify to the generality of the proposed model.

The proposed model describes thermal properties of investigated device at the steady state only. Currently the authors are conducting works the aim of which is to take into account in the presented model: thermal inertia, sizes and the shape of the heat-sink or printed circuit boards and the influence of the case of the device on efficiency of heat dissipation in the examined transistor.

6. References

[1] Górecki K and Zarębski J 2013 Metodyka formułowania wielodrogowego nieliniowego skupionego modelu termicznego elementu półprzewodnikowego *Elektronika* **54** (1) 22-25

[2] Górecki K and Zarębski J 2012 Paths of the heat flow from semiconductor devices to surrounding *Proceedings of the 19th Internationa Conference Mixed Design of Integrated Circuits and Systems MIXDES* 313-318

[3] Górecki K and Zarębski J 2010 Nonlinear compact thermal model of power semiconductor devices *IEEE Trans. on Components and Packaging Technologies* **33** (3) 643-647

[4] D'Alessandro V and Rinaldi N 2002 A critical review of thermal models for electro-thermal simulation *Solid-State Electronics* **46** (4) 487-496

[5] Szekely V 1997 A new evaluation method of thermal transient measurement results *Microelectronic Journal* **28** (3) 277-292

[6] Bagnoli PE, Casarosa C, Ciampi M and Dallago E 1998 Thermal resistance analysis by induced transient (TRAIT) method for power electronic devices thermal characterization *IEEE Trans. on Power Electronics, I. Fundamentals and Theory* **13** (6) 1208-19

[7] Oettinger FF and Blackburn DL 1990 Semiconductor measurement technology: thermal resistance measurements *U. S. Department of Commerce* **NIST/SP-400/86**

[8] Zarębski J and Górecki K 2007 A new measuring method of the thermal resistance of silicon p-n diodes *IEEE Trans. on Instrumentation and Measurement* **56** (6) 2788-2794

[9] Zarębski J and Górecki K 2007 A new method for the measurement of the thermal resistance of the monolithic switched regulator LT1073 *IEEE Trans. on Instrumentation and Measurement* **56** (5) 2101-2104

[10] Zarębski J and Górecki K 2007 A method of measuring the transient thermal impedance of monolithic bipolar switched regulators *IEEE Trans. on Components and Packaging Technologies* **30** (4) 627 – 631

[11] Zarębski J and Górecki K 1995 A method of the BJT transient thermal impedance measurement with the double junction calibration *Proceedings of 11-th IEEE Semiconductor Thermal Measurement and Management Symposium (SEMI-THERM)* 80-82

[12] Górecki K and Zarębski J 2009 Badanie wpływu wybranych czynników na parametry cieplne tranzystorów mocy MOS *Przegląd Elektrotechniczny* **85** (4) 159-164

[13] Górecki K and Zarębski J 2013 Semiconductor component thermal model taking into account non-linearity and multihipathing of the cooling system *Proceedings of Microtherm 2013 Microtechnology and Thermal Problems in Electronics* 71-77

[14] Zarębski J and Górecki K 2008 A method of the thermal resistance measurements of semiconductor devices with p-n junction *Measurement* **41** (3) 259-265

[15] Prasher R 2006 Thermal interface materials: historical perspective, status, and future directions *Proceedings of the IEEE* **94** (8) 1571-1586

[16] Padram M and Nazarian S 2006 Thermal modeling, analysis, and management in VLSI circuits: principles and methods *Proceedings of the IEEE* **94** (8) 1487-1501

[17] Bar-Cohen A, Arik M and Ohadi M 2006 Direct liquid cooling of high flux micro and nano electronic components *Proceedings of the IEEE* **94** (8) 1549-1570

[18] Sharp J, Bierschenk J and Lyon HB 2006 Overview of solid-state thermoelectric refrigerators and possible applications to on-chip thermal management *Proceedings of the IEEE* **94** (8) 1602-

1612
[19] Garimella SV, Singhal V and Liu D 2006 On-chip thermal management with microchannel heat sinks and integrated micropumps *Proceedings of the IEEE* **94** (8) 1534-1548
[20] Blackburn DL and Oettinger FF 1976 Transient thermal response measurements of power transistors *IEEE Trans. on Industrial Electronics and Control Instrumentation*, **IECI-22** (2) 134-141
[21] Zarębski J, Górecki K and Stepowicz WJ 1997 Wpływ sposobu zamocowania tranzystora na jego rezystancję termiczną *VI Konferencja Naukowa „Technologia Elektronowa" ELTE'97* **1** 708-712
[22] Górecki K, Zarębski J and Stepowicz WJ 2005 Wpływ wybranych czynników na parametry termiczne przyrządów półprzewodnikowych *Elektronika* **46** (10) 18-20
[23] 2002 IRF530 Data sheet *Fairchild Semiconductor* http://www.datasheetcatalog.org/datasheet/fairchild/IRF530.pdf

Acknowledgements
This project is financed from the funds of National Science Centre, which were awarded on the basis of the decision number DEC-2011/01/B/ST7/06740.

Journal of Physics: Conference Series **494** (2014) 012009 doi:10.1088/1742-6596/494/1/012009

Heat Transfer Enhancement by Finned Heat Sinks with Micro-structured Roughness

L Ventola[1], E Chiavazzo[1], F Calignano[2], D Manfredi[2], P Asinari[1]
[1] Multi-Scale Modeling Laboratory, Energy Dept., Politecnico di Torino, Corso Duca degli Abruzzi 24,Turin, Italy
[2] Center for Space Human Robotics @Polito, Istituto Italiano di Tecnologia, Corso Trento 21, Turin, Italy

Email: pietro.asinari@polito.it

Abstract. We investigated the benefits of micro-structured roughness on heat transfer performance of heat sinks, cooled by forced air. Heat sinks in aluminum alloy by direct metal laser sintering (DMLS) manufacturing technique were fabricated; values of the average surface roughness R_a from 1 to 25 microns (standard milling leads to roughness around 1 micron) under turbulent regimes (Reynolds number based on heating edge from 3000 to 17000) have been explored. An enhancement of 50% in thermal performances with regards to standard manufacturing was observed. This may open the way for huge boost in the technology of electronic cooling by DMLS.

1. Introduction

The ability of electronic devices to dissipate heat is an important aspect in electronics. Electronic cooling represents a commercial sector where a variety of solutions has been explored [1]. Skipping the very little performative solutions based on natural convection, thermal management of most of electronic devices, desktop and notebook computers to name a few, is based on forced convection of air as refrigerant, and usually rely on chip-attached or adhesively bonded extruded aluminum heat sinks [1, 2].

Concerning cooling of high power devices as high performance supercomputers, high thermal fluxes have already driven the development of liquid-based cooling loops. However, despite higher heat fluxes expected by these solutions, it is difficult to imagine a wide spread of the latter technology in notebook computers, which will remain dominated by cooling based on forced air convection.
Thermal performance of the air cooled heat sinks must be further improved due to ever-increasing thermal challenges arising in next-generation electronics systems [3]. Hence, in the present paper, we focused on forced (single-phase) convection only.

In particular we investigated the capability of direct metal laser sintering (DMLS) manufacturing technique to produce micro-structured rough surface heat sinks [4].

Usually heat sinks have been produced by milling manufacturing technique; surface roughness of those heat sinks is very low, around 1 micron. Through DMLS we are able to produce heat sinks whose surface roughness can vary over a wide range of values. This point is very important because the roughness strongly influences the convective heat transfer. The ability of roughness to enhance heat transfer in fully developed flow has been studied for many years. Theory and experimental correlations have been developed for this phenomenon. Early works focused on close packed, granular

Journal of Physics: Conference Series **494** (2014) 012009 doi:10.1088/1742-6596/494/1/012009

surface roughness [5, 6]. On the other hand, artificial-roughness-based solutions designed to enhance heat transfer (e.g. repeated ribs) have been studied [7, 8, 9]. In the latter cases, usually geometrical structured patterns are designed on the surface experiencing convective heat transfer. Recently it has become evident that the roughness has an even greater impact at the leading edge (or entrance region) of heat transfer surfaces, where the thermo-fluid dynamic structures (namely velocity and thermal boundary layer) start growing and they are very thin; thus they are more sensitive to structures that modify the flow field. This is the case of electronic cooling applications, because in this field geometric dimensions involved are usually very small and therefore the fraction of heat transfer at leading edge is far from negligible. Therefore we expect heat sinks produced by DMLS have better performance in heat dissipation than traditional those produced by milling.

In this work, through an experimental analysis in a wind tunnel, we compared the thermal performance of surfaces obtained by traditional milling to those obtained by DMLS. In particular two geometry will be tested, a flat heat sink and a finned heat sink (see figure 1).
For each setup, the average convective heat transfer coefficient h of two samples (with regards to the bare surface exposed to the air) was measured: A milled copper sample and an aluminum alloy (AlSi10Mg) sample produced by DMLS.

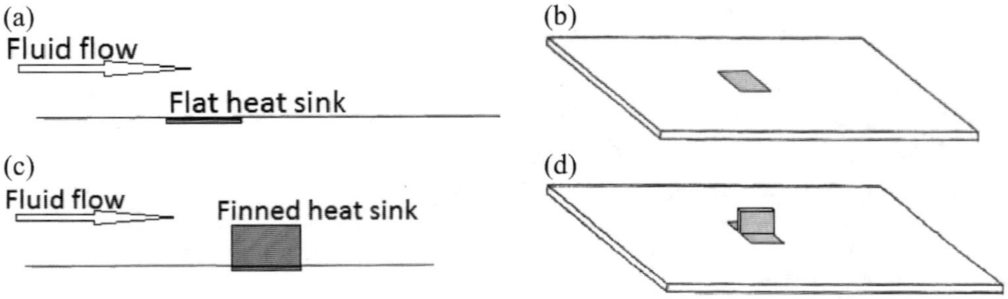

Figure 1. Longitudinal view of (a) flat and (c) finned heat sink; isometric view of (b) flat and (d) finned heat sink.

2. Direct Metal Laser Sintering process

The DMLS additive manufacturing process involves the use of a 3D CAD model of the part to product whereby a .stl file is created; then the required support structures are generated with the help of a dedicated software like Magics (from Materialise). The main functions of the supports structures are to hold unsupported geometries in place and to prevent the distortion of the part during fabrication, to fix the part to the building platform and to conduct excess heat away from the part. The complete built file, including parts and supports, is then sliced into layers with a chosen thickness (30 μm) and sent to the DMLS software. In an early stage the powders are sieved and put inside the dispenser in the building chamber filled with inert gas (Ar) in order to reach a level of oxygen lower than 0.1%. Once the process has started, the metallic powder is deposited with a stainless steel recoater blade on a building platform of similar material.

In this study a DMLS machine from EOS M270 Xtended machine was employed. This machine uses a 200W Ybfiber continuous laser beam with the wavelength of 1060 – 1100 nm. The focused diameter of laser beam is 0.1 mm. The maximum scan speed is 7000 mm/s. The building volume of the machine is 250 mm x 250 mm x 215 mm. The high-powered Yb-fiber optic laser fully melts the powder particles of the sliced .stl file at each layer. The substrate platform then drops one layer thickness in the z axis before the material is recoated, and the process is repeated until the entire build is complete. At the end, the building platform with the parts is removed from the building chamber and it is put into a furnace for a stress relieving treatment: for the Al alloy employed in this work this means 2 hours at 300°C. Only after this the supports can be removed and the parts detached.

The part is fabricated with different exposure DMLS process parameters for core, downskin, upskin and contour respectively [4]. In detail, downskin refers to the first two layers at the base of the part, while upskin refers to the last three layers at the top of the part. Core represents the remaining layers between the downskin and the upskin of the part. For each layer, after the hatching of the core region is completed, contour exposure is applied with lower laser power in order to improve the surface finish of the sintered part. The scan lines of every layer within the core region are rotated by 67° with respect to the scan lines of the preceding layer, as illustrated in a previous study [10].

Starting from the results obtained on the optimization of process parameters on surface finish of AlSiMg sample produced by DMLS, values which can modify and increase the surface roughness were chosen and two samples were manufactured.

Two samples produced by DMLS are analysed in this study. First sample dimensions are 11.1mm x 11.1mm x 5mm. The upper face is the surface exposed to the air flow, and the one on which the convective heat transfer takes place, while the other surfaces are isothermal. Second sample dimensions are 11.1mm x 11.1mm x 5mm, with a fin built on the upper face. Fin measures 11.1mm x 10mm x 2mm. In figure. 2 are shown the flat and finned heat sink obtained by DMLS and the corresponding copper ones. The samples were characterized by a 3D optical scanner ATOS Compact Scan 2M (GOM GmbH), and a numerical reconstruction of the flat heat sink surface is shown in figure 3. Surface roughness of copper samples is 1 micron, while surface roughness of flat and finned heat sink is 25 and 18 micron respectively.

Figure 2. Picture of milled copper and DMLS Aluminum alloy heat sinks (a) flat (b) finned.

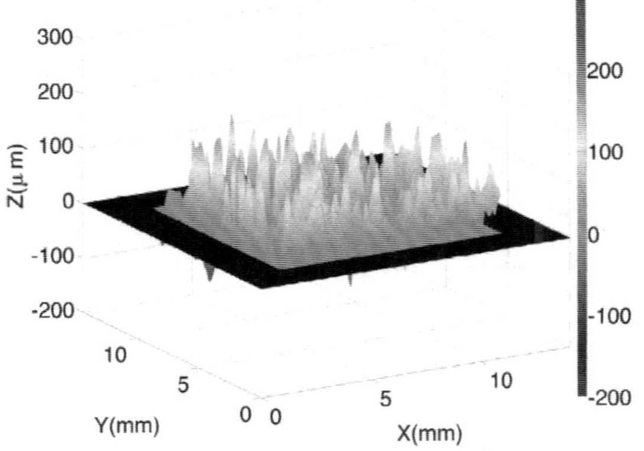

Figure 3. Numerical reconstruction of DMLS flat heat sink surface: z-axis has been magnified to show the nature of roughness obtained by DMLS process.

3. Experimental setup and procedure

We developed a sensor for the measurement of convective heat transfer coefficient. This sensor is based on the key idea of using a thermal guard. A heater in the bottom of the sample generate a controlled heat flux. The sample is surrounded by temperature controlled "guard" made of copper.This guard is kept at the same temperature of the sample, to force the heat to flows into the sample only.

This ensures that the heat flux through the sample is essentially one-dimensional, going from the heater to the air flow passing through the sample itself. Temperature are measured by thermocouples in the sample and in the upstream and downstream (referring to fluid flow) part of the guard. The upstream and downstream part of the guard experience different values of convective heat transfer coefficient, due to leading edge effect. Hence, the guard will be slightly non isothermal, and some (undesired) temperature gradient between the sample and the guard (of the order of 0.2 K) could arise. For this reason a thermal insulating shield made of Teflon is placed between sample and guard. Its function is to minimize parasite heat flux due to residual temperature gradients between sample and guard, In order to make them negligible.

The main portion of the power P provided by the heater to the sample is transferred into the air flow by convective heat transfer, while a minor, but far from negligible, portion of P is transferred to the wind channel walls through radiative heat transfer. Denoting by $Q_c$and Q_r the convective heat flux and the radiative heat flux respectively, the former could be calculated using the following equation:

$$Q_c = P - Q_r = P - A\epsilon\sigma(T_s^4 - T_w^4) \quad (1)$$

where A is the sample area, T_s is the sample temperature, T_w is the wind channel temperature, $\sigma = 5.67\ 10^{-8}\ W/m^2/K^4$ is the Stefan-Boltzmann constant, ϵ is the sample emissivity.

The sample emissivity has been estimated using an Infra-Red thermal camera. Values of 0.28 and 0.20 have been measured for Copper and aluminum alloy sample surface emissivity respectively.

The average temperature values of air and wind tunnel walls are 300 K and 299 K respectively. Their values are quite the same for all the tests. On the other hand $T_s$assumes very different values in each test, varying in the range of 307 ÷ 335 K. Consequently also Q_r assumes different values in each test. However, the maximum Q_r experienced in the tests is less than 6% of the overall power P provided by the heater.

The flow loop of the experimental system is shown schematically in figure 4. The sensor was mounted on one side of a horizontally oriented rectangular flow channel, which realizes a small open-loop wind tunnel. The channel has a smooth inner surface, a section of 228mm x 158mm (hydraulic diameter $D_h =$ 187mm) and an entrance length of 5m (corresponding roughly to 26 hydraulic diameters). At the end of the channel, downstream from the test section, a vein anemometer was installed for measuring the axial velocity. The inner wall and air temperature are measured at the same location where the anemometer is located.

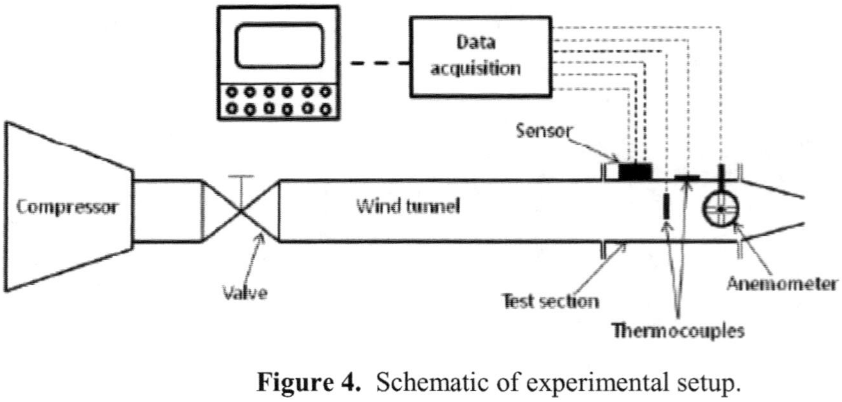

Figure 4. Schematic of experimental setup.

4. Results

Two different test cases were analyzed: case A and case B refer to the flat and finned heat sinks, respectively. For each test case, the average convective heat transfer coefficients (with regards to the surface exposed to the air flow) of the two samples was compared: A milled copper sample and an aluminum alloy (AlSi10Mg) sample produced by DMLS. Convective heat transfer coefficient was evalued for different velocities, in order to explore the nature of the heat transfer phenomenon in a wide range of Reynolds numbers. To each value of mean velocity inside the wind tunnel u we associated the following dimensionless Reynolds number Re_L, calculated according to the formula:

$$Re_L = \frac{uL}{\nu} \tag{2}$$

with $\nu = 1.493 \,\text{x}10^{-5}\, m^2/s$ being kinematic viscosity of air and $L = 20\, mm$ being the characteristic length of heat source. Conversely, the average convective heat transfer coefficient h was expressed in terms of dimensionless number $Nu/Pr^{1/3}$, whereas:

$$Nu = h\frac{L}{k} \tag{3}$$

is the dimensionless Nusselt number and

$$Pr = \frac{\nu}{\alpha} \tag{4}$$

is dimensionless Prandtl number with $\alpha = 2.224\text{x}\,10^{-5}$ being the thermal diffusivity of air. To compare DMLS samples and milled samples we introduced the thermal performance enhancement defined as:

$$En = \frac{h_{DMLS} - h_{milled}}{h_{DMLS}} 100 \;\; [\%] \tag{5}$$

4.1. Test case A: Flat heat sinks

We ran 40 tests for DMLS sample and 15 tests for milled copper sample, and extrapolated from experimental data the following fitting curves:

$$Nu/Pr^{1/3} = 0.082\, {Re_L}^{0.669} \tag{6}$$

for milled sample ($Ra = 1\mu$)

$$Nu/Pr^{1/3} = 0.093\, {Re_L}^{0.732} \tag{7}$$

for DMLS sample ($Ra = 25\mu$). In figure 5 experimental points and interpolating curves are shown. Enhancement obtained is quite constant over the whole range of explored Re_L, with a minimum and maximum value of 47.7% and 56.4% respectively.

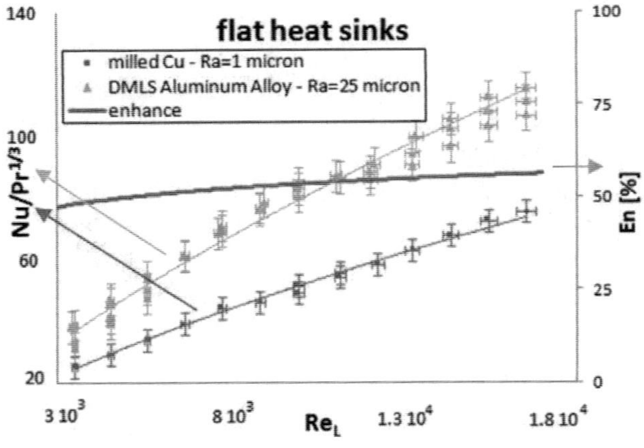

Figure 5. Flat heat sinks, experimental results. Red and green data report values of $Nu/Pr^{1/3}$ quantity, so they refer to the left axis of ordinates. Blue data showing the trend of thermal performance enhancement refers to the right axis of ordinates.

4.2. Test case B: Finned heat sinks

Thermal conductivity k of copper is more than AlSi10Mg one.This gives no effect on flat heat sinks, while in case of finned samples it is necessary to check if fin efficiency, due to conduction into the fin, may spoil the heat transfer enhancement or not. Fin efficiency $\eta$is calculated by means of the following formula [2]:

$$\eta = \frac{\tanh(m*L_{fin})}{m*L_{fin}} \quad (8)$$

$$m = \left(\frac{2h}{t*k}\right)^{-1} \quad (9)$$

where $L_{fin} = 10\,mm$ is fin length, while $t = 2\,mm$ is fin thickness. $k_{Cu} = 388\,\frac{W}{m^2K}$ while $k_{\text{AlSi10Mg}} = 150\,\frac{W}{m^2K}$. To calculate fin efficiencies in equation (9) we chose the maximum value of h we obtained in the experiment. Therefore, considering the maximum value of h, we calculated the minimum value of the efficiencies of the fin and the maximum gap between the efficiencies of the two finned samples ($\Delta\eta$):

$$\eta_{copper} = 98.47\,\%;\ \eta_{Aluminumalloy} = 96.18\,\%;\ \Delta\eta = 2.29\,\%$$

This choice of h allow us to consider the most conservative case. Since $\Delta\eta$ is very small (2.29%), we can assume that the conductivity disparity of the two materials under study does not affect heat transfer performance of samples. We ran 13 tests with both DMLS and milled sample, then we extrapolated from experimental data the following fitting curves:

$$Nu/Pr^{1/3} = 0.124\,{Re_L}^{0.722} \text{ for milled sample } (R_a = 1\mu) \quad (10)$$

$$Nu/Pr^{1/3} = 0.153\,{Re_L}^{0.720} \text{ for DMLS sample } (R_a = 18\mu) \quad (11)$$

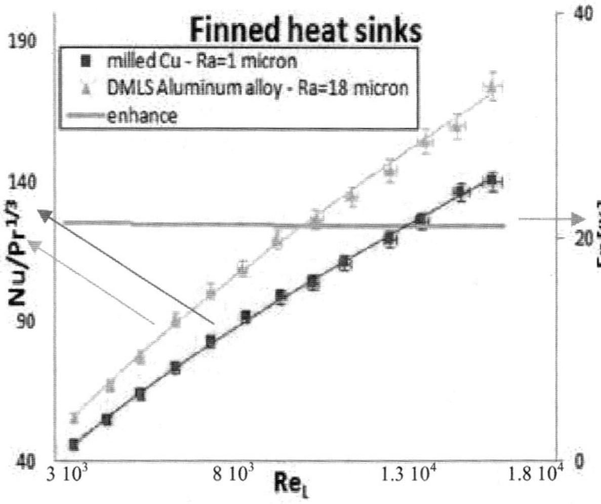

Figure 6. Finned heat sinks, experimental results.

In figure 6 we report the experimental results, the fitting curves and the thermal performance enhancement. Clearly the latter is nearly constant all over the explored Reynolds number range, and it amounts to 20%. It is very important to point out that fin efficiency is a parameter concerning conduction into the material only. Therefore it is plausible that the values of convective heat transfer coefficient measured on two samples are very different although the fin efficiency of the two samples is approximately equal.

4.3. Comments

It is important to know the range of velocity and Reynolds number values for industrial application heat sinks. Comparing the latter with the range of the Reynolds number explored in our experimental campaign, it is possible to predict if the enhancement in heat transfer showed in our study will be achieved also when DMLS rough heat sinks works under fluid dynamics conditions typical of industrial applications.

Length of flat heat sinks or finned heat sinks basis plate usually vary from 2 to 15 cm, depending on the heat flux to dissipate. On the other hand velocity of cooling air for finned heat sinks electronic devices usually ranges from 1 to 5 m/s [11]. Consequently, according to equation (2) , in the most of industrial electronic cooling devices Re_L varies from $1.3\ x\ 10^3$ to $5.0\ x\ 10^4$. Hence we can argue that the experimental results shown in this work have been ran in a range of fluid dynamic conditions which are of interest for electronic cooling devices for industrial applications. Consequently it is possible to suppose the enhancement in heat transfer through DMLS manufacturing showed in the present work will be achieved also if DMLS rough heat sinks will be applied to common industrial electronic cooling devices.

It is appropriate to consider the problem of pressure losses in the electronic cooling device air circuits. Usually heat sinks are designed in order to obtain an optimum trade-off between heat transfer performance and pressure losses. In this study pressure losses associated to tested heat sinks have not been characterized. However it is reasonable to expect an increase in pressure losses when a rough heat sink is used instead of a smooth one. In spite of the presence of heat sinks is a source of pressure losses, it should be taken into account that a big portion of pressure losses in air circuits are due to other component rather than heat sinks. Pressure losses due to other component remain approximately constant, while varying type of heat sink. As usually done, pressure losses can be subdivided in distributed pressure losses along channel length and concentrated pressure losses due to fan intake, bends and restriction along the circuit, to name a few. Reynolds analogy states that in first

Journal of Physics: Conference Series **494** (2014) 012009 doi:10.1088/1742-6596/494/1/012009

approximation $Nu_{HS} = f_{HS}$, where Nu_{HS} and f_{HS} are the dimensionless heat and momentum transfer coefficient on the heat sinks respectively.

Increasing Nu_{HS} proportionally increase the dissipating heat flux, being heat sinks the component of the circuit which mainly experience heat transfer phenomenon, whilst increasing f_{HS} proportionally increase only pressure losses associated to heat sinks, while pressure losses due to other component remain constant. Hence we can suppose that, when a rough heat sink is used instead of a smooth one, the benefit in enhancing heat transfer should be more effective than the negative effects of increasing pressure losses.

In this work, the effect of DMLS process surface roughness on thermal performance of air cooled heat sinks has been investigated. Nevertheless there are several technologies for electronic cooling, other than air cooled heat sinks. For example, confined jet impingement, heat pipe and so on. It is interesting to compare the presented technique with others, in order to prove the relevance of the obtained results in the electronic cooling research field. In general, air cooled heat sinks are designed to take advantage of extended surfaces, for example plate fins or pin fins. Jet impingement can also be incorporated into the design to further enhance the heat transfer. Because available space is limited and low power consumption is desired, the heat sink performance becomes the focus of many studies. In fact the latter is a widespread method in computers and CPU cooling due to its low price, effectiveness and reliability [12]. Recently, the class of techniques based on liquid or two-phase coolant fluids, as heat pipes to name one, became very attractive. Heat transfer mechanism in heat pipe combines the principles of both thermal conductivity and phase transition to tremendously increase heat transfer rate [12]. Despite this class of techniques allow to dissipate very high heat fluxes (compared to air cooled heat sinks) they are still relatively expensive. Hence, despite the common belief that the limits of air-cooling have been reached, yet a number of interesting approaches can be found in the literature. For example in [13] an innovative transient thermal management solutions that utilize PCMs (phase change materials) to design hybrid PCM-based air cooled heat sinks is presented. It provides a solution to deal with peak transient heat loads in portable applications exploiting the ability of PCMs to store energy in the form of latent heat.

We can reasonably suppose that a lot of electronic applications (like notebooks) will remain dominated by cooling based on forced air convection in the next future. Consequently it is important to find techniques, like the one exposed in this work, capable to improve thermal performance of air cooled devices to face the ever-increasing thermal challenges arising in next-generation electronics systems.

5. Conclusions

In this study average convective heat transfer coefficient of different heat sinks was measured using air as refrigerant fluid in turbulent regime. Copper heat sinks produced by traditional milling were compared to aluminum alloy (AlSi10Mg) heat sinks produced by DMLS. Effects of micro-structured roughness on thermal performance were shown. These are very huge because little geometrical dimensions are involved in heat transfer phenomenon (see leading edge), as usual in electronic cooling. Our results offer an evidence of the possible impact of DMLS on electronic cooling since a 50% and 20% enhancement (compared to milled samples) is observed for flat and finned heat sinks, respectively. We also show that the lower thermal conductivity of aluminum alloy plays a negligible role on heat transfer performances in our setup. Enhancement on flat heat sinks is larger than finned sinks, because average roughness of the former is higher than that of the latter. Building rough fins is not as easy as doing rough flat surfaces, but in future studies this will be one of the first improvements that will be pursued from the point of view of DLMS machining. These results open the way for a huge boost in the technology of electronic cooling by DMLS. As an outlook, based on the present results, we plan further studies focusing on investigation and optimization of micro-structured surfaces obtained by DMLS with regards to heat transfer.

6. References

[1] Sergent J and Krum A 1998 *Thermal Management Handbook for Electronic Assemblies* (New York: McGraw-Hill)

[2] Bejan A and Kraus A D 2003 *Heat Transfer Handbook* (Hoboken,NJ: John Wiley & Sons)

[3] Garimella S V, Fleischer A S, Murthy J Y, Keshavarzi A, Prasher R, Patel C, Bhavnani S H, Venkatasubramanian R, Mahajan R, Joshi Y, Sammakia B, Myers B A, Chorosinski L, Baelmans M, Sathyamurthy P and Raad P E 2008 Thermal challenges in next-generation electronic systems *IEEE Trans. Compon. Packag. Manuf. Technol.* **31** 801-815.

[4] Calignano F, Manfredi D, Ambrosio E P,Iuliano L and Fino P 2013 Influence of process parameters on surface roughness of aluminum parts produced by DMLS *Int. J. Adv. Manuf. Technol.* **67** 2743-51

[5] Nikuradse J 1950 Laws of flow in rough pipes *N.A.C.A. tech. mem.* **1292**

[6] Dipprey D F and Sabersky R H 1963 Heat and momentum transfer in smooth and rough tubes at various Prandtl numbers *Int. J. Heat Mass Transfer.* **6** 329-332.

[7] Webb R L, Eckert E R G and Goldstein R J 1971 Heat transfer and friction in tubes with repeated-rib roughness *Int. J. Heat Mass Transfer.* **14** 601-617

[8] Li X W, Meng J A and Li Z X 2007 Experimental study of single-phase pressure drop and heat transfer in a micro-fin tube *Exp. Therm Fluid Sci.* **32** 641-648

[9] Li X W, Meng J A and Li Z X 2011 Roughness enhanced mechanism for turbulent convective heat transfer *Int. J. Heat Mass Transfer* **54** 1775-81

[10] Manfredi D, Calignano F, Krishnan M, Canali R, Ambrosio E P and Atzeni E 2013 From powders to dense metal parts: characterization of a commercial AlSiMg alloy processed through direct metal laser sintering *Materials* **6** 856-869

[11] Shaukatullah M, Storr W R, Hansen B J and Gaynes M A 1996 Design and optimization of pin fin heat sinks for low velocity applications *IEEE Trans. Compon. Packag. Manuf. Technol.* **19** 486-494

[12] Gururatana S 2012 Heat transfer augmentation for electronic cooling *Am. J. Appl. Sci.* **9** 436-439

[13] Rodgers P, Eveloy V, and Pecht M G 2005 Limits of air-cooling: status and challenges *21st IEEE SEMI-THERM Symp.* **1** 116-124

Acknowledgments
Authors would like to acknowledge the THERMALSKIN project: Revolutionary surface coatings by carbon nanotubes for high heat transfer efficiency (FIRB 2010 – "Futuro in Ricerca", grant RBFR10VZUG).

Journal of Physics: Conference Series **494** (2014) 012010 doi:10.1088/1742-6596/494/1/012010

A System for Cooling Electronic Elements with an EHD Coolant Flow

M Tanski[1], M Kocik[1], R Barbucha[1], K Garasz[1], J Mizeraczyk[1,2], J Kraśniewski[3], M Oleksy[3], A Hapka[3] and W Janke[3]

[1] Centre for Plasma and Laser Engineering, The Szewalski Institute of Fluid Flow Machinery, Fiszera 14, 80-231 Gdańsk, Poland
[2] Department of Marine Electronics, Gdynia Maritime University, Morska 81-87, 81-225 Gdynia, Poland
[3] Department of Electronics and Computer Science, Technical University of Koszalin, Śniadeckich 2, 75-453 Koszalin, Poland

E-mail: tanski@imp.gda.pl

Abstract. A system for cooling electronic components where the liquid coolant flow is forced with ion-drag type EHD micropumps was tested. For tests we used isopropyl alcohol as the coolant and CSD02060 diodes in TO-220 packages as cooled electronic elements. We have studied thermal characteristics of diodes cooled with EHD flow in the function of a coolant flow rate. The transient thermal impedance of the CSD02060 diode cooled with 1.5 ml/min EHD flow was 7.8°C/W. Similar transient thermal impedance can be achieved by applying to the diode a large RAD-A6405A/150 heat sink. We found out that EHD pumps can be successfully applied for cooling electronic elements.

1. Introduction

As electronic elements become smaller and more power consuming, more problems with dispersing waste heat arise. Nowadays, traditional cooling techniques are no longer applicable in many fields of technology because of a large size of cooling systems and their low efficiency. Currently, lack of efficient technique of heat removal from electronic elements is one of major obstructions to further miniaturization of an electronic system in such applications as computers, biomedicine, automobile and aerospace technology [1]. Therefore, over the past few decades various cooling systems for electronic elements were proposed, for example: impinging air jets [2], thermoelectric coolers [3], direct immersion cooling [4], miniature heat-pipes [5] and liquid microflow cooling systems. Compared to other cooling techniques, liquid microflow systems seem to be an ideal cooling solution in applications where high efficiency and a small size of the cooling system are indispensable. In this cooling technique, liquid coolant is introduced close to the heat critical part of the electronic element by means of micropumps. Many pumping mechanisms for microflow cooling systems were tested [6]: from traditional mechanical pumps, through phase change pumps [7], magnetokinetic pumps [8], electroosmotic [9] and finally to electrohydrodynamic (EHD) pumps.

The advantages of the EHD pumps are: non-mechanical mechanism of pumping, relatively simple design, low vibration and noise emission, low power consumption and simple control of their efficiency by varying the applied voltage. In general, when electric field E is applied to a volume of dielectric liquid its driving force can be expressed as (1) [10].

$$F = qE - \frac{1}{2}E^2\nabla\varepsilon + \frac{1}{2}\nabla\left[E^2\left(\frac{\partial\varepsilon}{\partial\rho}\right)_T\rho\right] \quad (1)$$

where q is the free space charge density, ε is the liquid permittivity and ρ is its mass density. The first term of equation (1) represents the Coulomb force which is main driving force in most types of EHD pumps. The second term is dielectrophoretic force caused by the gradient of liquid permittivity. The last term is electrostrictive force. There are three basic mechanisms of EHD pumping in which Coulomb force is the driving force: induction pumping, conduction pumping and ion-drag (injection) pumping [11]. In our system for cooling electronic elements we applied a miniature ion-drag type EHD pumps. The basic mechanism of ion-drag pumping was described by Stuetzer [12] and Pickard [13] in 1960's and was investigated by many groups since then. During the ion-drug pumping free charges are injected into the dielectric liquid from the positive electrode (called emitter). The charges can be injected into the liquid by either a field ionization or field emission process. Next, injected charges are dragged by the Coulomb force towards another electrode (called collector) in high electric field created between the electrodes. Through the friction, a part of kinetic energy of the dragged charges is transferred to adjacent liquid what sets it into motion.

In this paper we present the design and test results of the system for cooling electronic elements in which liquid coolant flow is forced by the miniature, ion-drag type EHD pumps.

2. Experimental set-up

The experimental set-up of the system for cooling electronic elements is presented in figure 1. It consisted of an electronic element (diode Cree CSD02060 SiC in TO-220 package) with integrated U-shaped heat exchanger, set of three EHD pumps (parallel or serial connected) and reservoir of the liquid coolant with traditional heat-sink. Heat exchanger integrated with the diode was done by manufacturing in package of the diode a pattern of microchannels by means of laser micromachining technique. The EHD pumps force the coolant flow, in the closed-loop, from the reservoir to the heat exchanger and back to the reservoir.

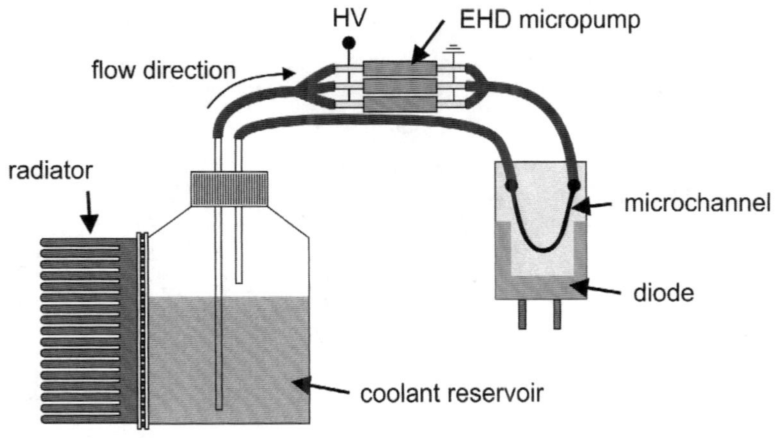

Figure 1. Cooling system for TO-220 packed diode with an EHD-induced coolant flow.

2.1. EHD micropump

The ion-drag type EHD micropump (figure 2) is composed of a ceramic (Al_2O_3) pipe and two stainless-steel electrodes inserted into the ceramic pipe. Each electrode is a 10 mm pipe of external and internal diameter of 0.8 mm and 0.5 mm, respectively. Distance between electrodes was 3 mm. The positive high DC voltage was applied to one electrode while the other electrode was grounded.

Spellman SL300 power supply was used as high voltage source. The pumping effect was observed in the direction from the HV electrode (emitter) towards the grounded electrode (collector). Pure isopropyl alcohol $(CH_3)_2CHOH$ was used as a coolant.

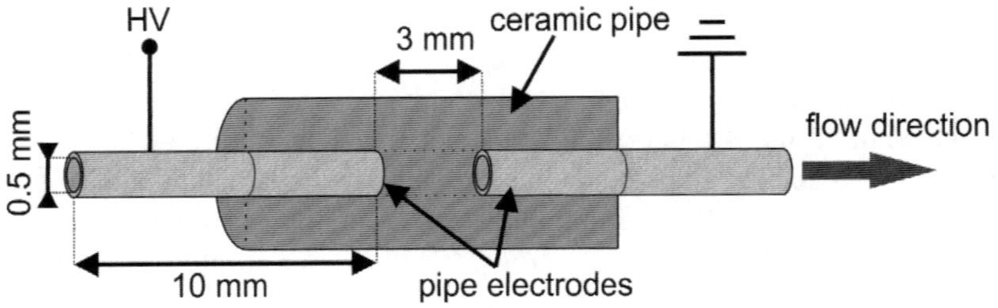

Figure 2. Schematics of EHD pump construction.

The maximum flow rate generated by a single EHD pump was 0.5 ml/min and the maximal pressure was 500 Pa for applied voltage of 12 kV. For higher voltages we observed formation of the gas bubbles in the pumped coolant and in consequence significant drop in both flow rate and pressure. For applied voltage of 12 kV the measured current was 20μA, thus EHD pump power consumption was 240 mW.

2.2. Microchannels fabrication

In order to make a heat exchanger integrated with the diode we carried out a pattern of microchannels in the metal part of diode package by means of laser micromachining technique, so that the coolant could be introduced close to a heat-critical area of the diode. Geometry of a typical U-shaped microchannel is presented in figure 1. Microchannel was both 500 μm wide and deep and 2 cm long. The microchannel was covered with transparent polidimetylosiloksan (PDMS) resin and thin glass plate. Transparent resin and glass plates were used in order to verify that no gas bubbles were generated in the coolant during the cooling process. The cross-section of complete heat exchanger is presented in figure 3.

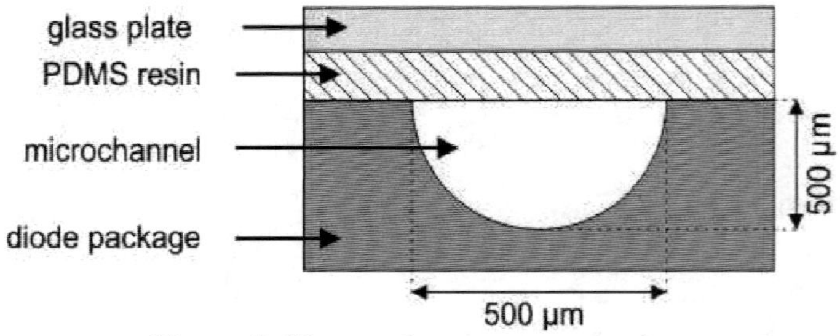

Figure 3. Heat exchanger cross-section.

3. Results

3.1. Micro-PIV measurements

To measure the EHD flow profile across the microchannel we propose to use the micro PIV technique. The micro-PIV stand included a dual pulsed Nd:YAG laser (532 nm, 15 ns pulse duration, 20 mJ pulse energy), a microscope NIKON Eclipse with a Kodak Megapixel CCD camera and Dantec Flow Manager software on a PC. The working liquid contained fluorescent seeding particles (polyamide spheres of diameter 1 and 10 um with Rhodamine B). During a single measurement, particles were flashed with the laser twice, making the particle emit the fluorescence light. After each flash an image of fluorescent light was taken. The resulting pictures were analyzed to determine the movement of particles. Once the time between laser pulses was known, a flow velocity map could be determined.

Journal of Physics: Conference Series **494** (2014) 012010 doi:10.1088/1742-6596/494/1/012010

A series of such measurements was performed for each observation area along the microchannel and an averaged flow velocity profile and a flow rate was calculated.
The maximum flow rate generated by the three EHD micropumps connected in series and in parallel was measured (figure 4). In case of parallel connection the total flow rate was 1.6 ml/min (flow rate generated by single EHD pump was 0.5 ml/min). On the other hand, in the case of serial connection of three EHD pumps the resulted flow rate was only 0.8 ml/min. Therefore in the final version of the cooling system EHD pumps were connected in parallel.

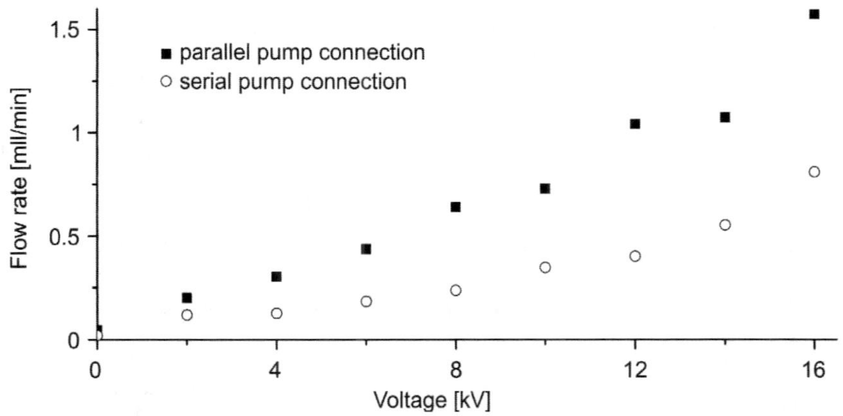

Figure 4. Coolant flow rate in the function of applied voltage for parallel and serial connection of the three EHD pumps.

3.2. Thermal measurements

To examine the performance of the cooling system a thermal impedance of the cooled CSD02060 diode was measured for various coolant flow rates. To obtain the transient thermal impedance first the cooling characteristics was taken. To this end an electric current was applied to the diode, heating it until the electrothermal steady state was reached. Next, the current was switched off and the diode started to cooled down to the ambient temperature. During this cooling process, the thermosensitive parameter of the diode was measured. This parameter was used to calculate the temperature surplus of a measured diode over the ambient temperature. Finally, transient thermal impedance curves were obtained by inversing the cooling curve and dividing the temperature surplus by the value of power used for heating the diode.

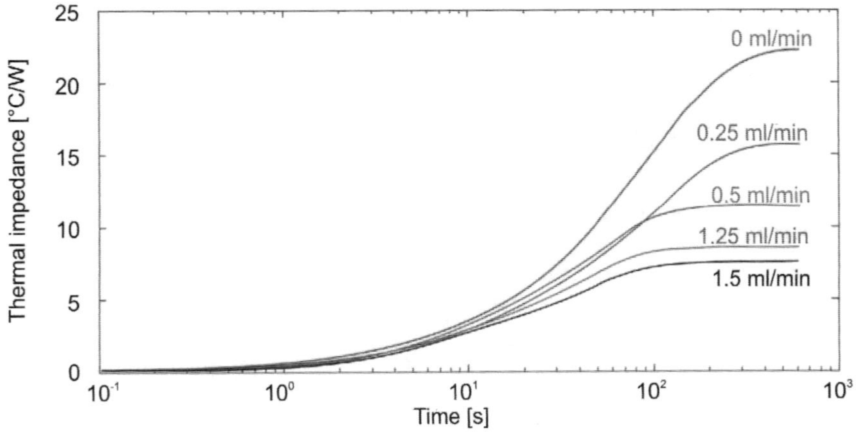

Figure 5. Thermal impedance of the CSD02060 diode.

4. Summary and conclusions

It was demonstrated that ion-drag type EHD pumps can be successfully applied for cooling electronic elements. We have studied the thermal characteristics of the CSD02060 diode cooled with an EHD flow in the function of a coolant flow rate. The transient thermal impedance of the CSD02060 diode cooled with 1.5 ml/min EHD flow was 7.8°C/W. Similar thermal resistance (6°C/W) can be achieved by applying a large RAD-A6405A/150 heat sink to the diode.

5. References

[1] Darabi J and Ekula K 2003 Development of a chip-integrated micro cooling device *IEEE Workshop on Embedded Systems Codsign* **34** 1067-74

[2] Wits W W, Vaneker T H J, Mannak J H and Legtenberg R 2009 Novel cooling strategy for electronic packages: Directly injected cooling *CIRP J. Manuf. Sci. Tech.* **1** 142-147

[3] Zhang H Y, Mui Y C and Tarin M 2010 Analysis of thermoelectric cooler performance for high power electronic packages *Appl. Therm. Eng.* **30** 561-568

[4] Warrier P, Sathyanarayana A, Patil D V, France S, Joshi Y and Teja A S 2012 Novel heat transfer fluids for direct immersion phase change cooling of electronic systems *Int. J. Heat Mass Transf.* **55** 3379-85

[5] Reay D and Kew P 2006 *Heat pipes. Theory, design and applications* (Oxford:Butterworth-Heinemann) p 319

[6] Singhal V, Garimella S V and Raman A 2004 Microscale pumping technologies for microchannel cooling systems *Appl. Mechanics Rev.* **57** 191–221

[7] Kuo S, Shih C, Chang C and Chen S 2013 Bubble pump in a closed-loop system for electronic cooling *Appl. Therm. Eng.* **51** 425-434

[8] Xuan Y and Liam W 2011 Electronic cooling using an automatic energy transport device based on thermomagnetic effect *Appl. Therm. Eng.* **31** 1487-94

[9] Berrouche Y, Avenas Y, Schaeffer C, Chang H and Wang P 2009 Design of a Porous Electroosmotic Pump Used in Power Electronic Cooling *Trans. Ind. Appl.* **45** 2073-79

[10] Nguyen N, Huang X and Chuan T K 2002 MEMS-Micropumps: A Review *J. Fluids Eng.* **124** 384-392

[11] Seyed-Yagoobi J 2005 Electrohydrodynamic pumping of dielectric liquids *J. Elect.* **63** 861-869

[12] Stuetzer O M 1959 Ion-drag pumping *J. Appl. Phys.* **31** 136-146

[13] Pickard W F 1963 Ion-drag pumping I theory *J. Appl. Phys.* **34** 246-250

Emission and Structure Varieties in ZnO:Ag Nanorods Obtained by Ultrasonic Spray Pyrolysis

E Velázquez Lozada[1*], T V Torchynska[2], J L Casas Espinola[2]
[1] ESIME – Instituto Politécnico Nacional, México D.F. 07738, México.
[2] ESFM – Instituto Politécnico Nacional, México D.F. 07738, México.

[*]E-mail: evlozada5@yahoo.com.mx

Abstract. Scanning electronic microscopy (SEM), X ray diffraction (XRD) and photoluminescence (PL) have been applied to the study of the structural and optical properties of ZnO nanocrystals prepared by the ultrasonic spray pyrolysis (USP) at different temperatures. The variation of temperatures and times at the growth of ZnO films permits modifying the ZnO phase from the amorphous to crystalline, to change the size of ZnO nanocrystals (NCs), as well as to vary their photoluminescence spectra. The study has revealed three types of PL bands in ZnO NCs: defect related emission, the near-band-edge (NBE) PL, related to the LO phonon replica of free exciton (FE) recombination, and FE second-order diffraction peaks. The PL bands related to the LO phonon replica of FE in PL spectra measured at room temperature testify on the high quality of ZnO films prepared by the USP technology.

1. Introduction

Nanocrystalline Zinc oxide (ZnO) with the band gap energy nearly 3.37 eV and high exciton binding energy (60 meV at 300K) has attracted great attention during the last two decades [1-3]. In addition the ZnO possesses a number of deep levels that emit in the visible range and, hence, can provide intrinsic "white" light emission. ZnO nanostructures are being investigated as promising candidates for different optoelectronic applications, such as: the non-linear optical devices [4], light-emitting devices [5-8], transparent electrodes for solar cells [9] and laser diodes [10], as well as for the excellent field emitters [11], electrochemical sensors and toxic gas sensors [12].

The control of ZnO defect structure in nanostructures is a necessary step in order to improve the device quality. Since the structural imperfections and defects generally deteriorate the exciton related recombination process, it is necessary to grow the high quality films for efficient light-emitting applications. The ultrasonic spray pyrolysis (USP) method offers many advantages such as easy compositional modifications; easy introducing the various functional groups, relatively low annealing temperatures and the possibility of coating deposition on a large area substrate[13, 14].

It will be interesting to study the optical emission of USP produced ZnO nanorods (NRs) doped by Ag in order to identify the best regimes for obtaining the bright emitting nanosystems.

2. Experimental details

ZnO:Ag thin solid films were prepared by the USP technique (Fig. 1) on the surface of soda-lime glass substrate for the two substrate temperatures (400 and 450° C) and different deposition times (Table 1). The design of deposition system is presented in figure 1[15]. This system includes a piezoelectric transducer operating at variable frequencies up to 1.2 MHz and the ultrasonic power of 120 W.

Table 1. Technological regimes and NR parameters from SEM images

Sample numbers	Growth temperature (°C)	Deposition time (min)	NR width (nm)
1	400	10	150-200
2	400	5	100-150
3	400	3	50-70
4	450	10	200-250
5	450	5	120-180
6	450	3	80-110

ZnO:Ag thin solid films were deposited from a 0.4 M solution of zinc (II) acetate [Zn(O2CCH3)2] (Alfa), dissolved in a mix of deionized water, acetic acid [CH3CO2H] (Baker), and methanol [CH3OH] (Baker) (100:100:800 volume proportion).

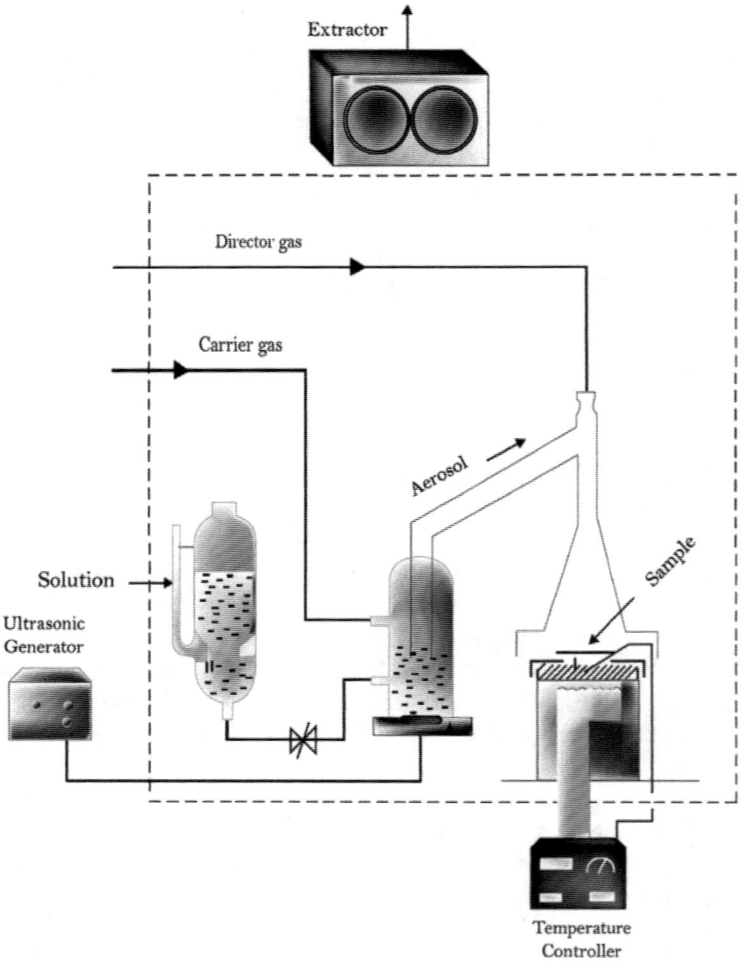

Figure 1. Schematic design of experimental setup used for the deposition of ZnO:Ag films by the ultrasonic spray pyrolysis method.

Separately, a 0.2 M solution of silver nitrate [Ag (NO_3)] (Baker) dissolved in a mix of deionized water and acetic acid [CH_3CO_2H] (Baker) (1:1 volume proportion) was prepared, in order to be used as a doping source. A constant [Ag]/[Zn] ratio of 2 at. % was applied at the ZnO Ag film preparation.

The morphology of ZnO:Ag films has been studied by the scanning electron microscopy (SEM) Dual Beam, FEI brand, model Quanta 3D FEG. PL spectra were measured at the excitation by a He-Cd laser with a wavelength of 325 nm and a beam power of 50 mW at 300K using a PL setup on a base of spectrometer SPEX500 described in [16, 17]. The crystal structure of ZnO:Ag films was investigated by the X-ray diffraction (XRD) method using the diffractometer Model XPERT MRD with the Pixel detector, three axis goniometry and parallel collimator with the resolution of 0.0001 degree. XRD beam was from the Cu source, Kα1 line λ=1.5406 Å, 45 kV, 40 mA and the angles used were from 20° to 80° with a step size of 0.05° and step time of 100 s.

3. Experimental results and discussion

SEM images of typical ZnO:Ag NRs obtained at the deposition times of 3 and 10 min for two substrate temperatures are presented in figure 2. It is clear that the ZnO NRs have the hexagonal cross section and the road orientation along the c axis. The cross section of ZnO NRs increases with the temperature and UPS durations (Table 1).

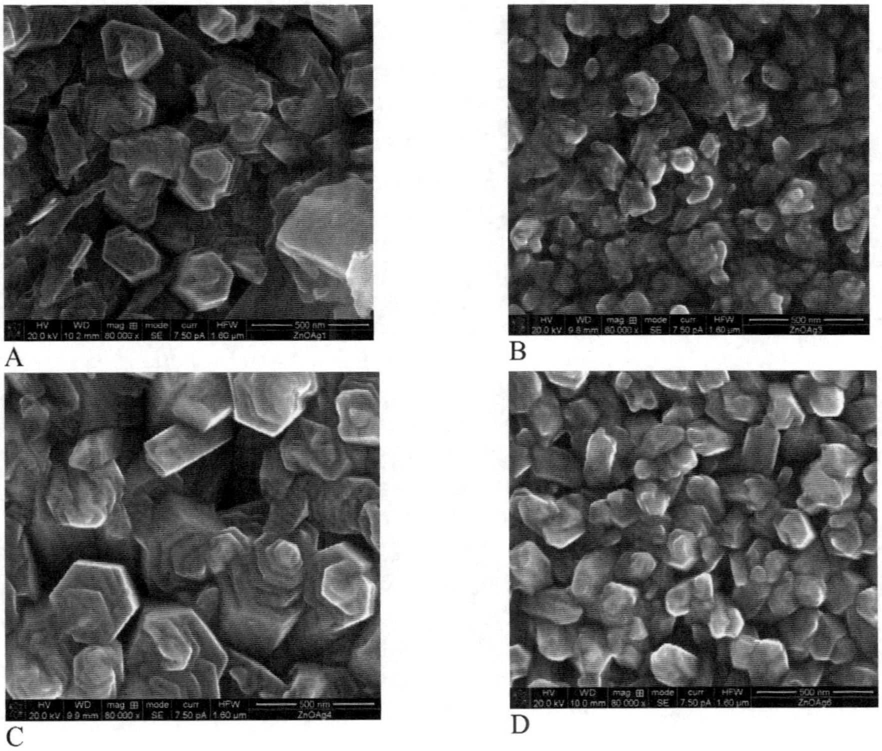

Figure 2. SEM images of the samples prepared at the substrate temperatures 400°C (a, b) and 450°C (c, d) and the durations of 10 (a, c) and 3 (b, d) min.

X-ray diffraction patterns of ZnO:Ag NRs obtained on the substrate with the temperature of 400°C and different deposition durations are shown in Fig. 3. At low deposition time (3min) the ZnO films showed an amorphous phase mainly with very small (100), (002) and (10 1) peaks. These peaks are the evidence of starting the conversion of an amorphous phase into a polycrystalline one. It was observed that with increasing the deposition duration from 3min to 10min, a set of new peaks appears, which correspond to the X- ray diffraction from the (100), (002), (101), (102), (110), (103) and (200) crystal planes with the angles 2θ equal to 31.770, 34.422, 36.253, 47.540, 56.604, 62.865 and 68.709 degrees, respectively, in the wurzite ZnO crystal structure with the hexagonal lattice parameters of a=3.2498A and c=5.2066A [18]. The (0 0 2) reflection peak is intense and sharper, as compared to the other peaks, indicating a preferential c-axis orientation of ZnO:Ag NRs.

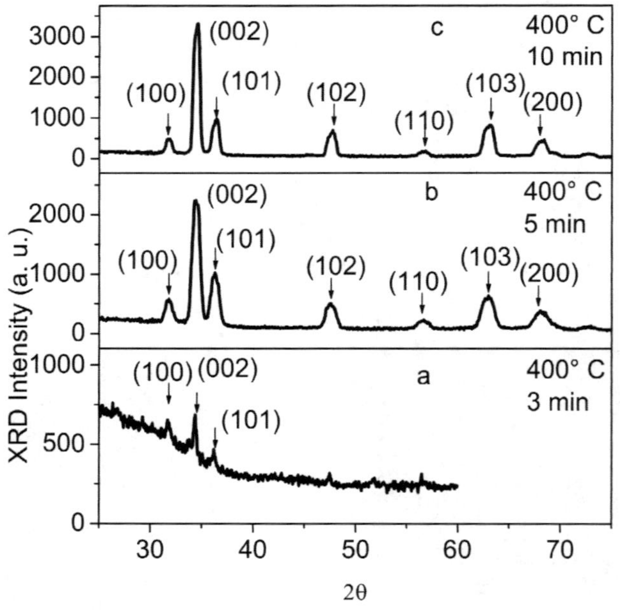

Figure 3. XRD diagrams of studied samples prepared on the substrate with T = 400° C at the deposition times of 3 (a), 5 (b) and 10 (c) min.

PL spectra of ZnO:Ag NRs are shown in figure 4. It is clear that PL spectra are complex and can be represented as a superposition of elementary PL bands with the peaks at: 3.14 eV (I), 2.00-2.70 eV (II) and 1.57 eV (III) (Fig.4). It is known that the UV-visible PL bands in ZnO owing to the near-band-edge (NBE) or exciton (I) and defect-related (II) recombination [19-22]. With increasing the substrate temperature the NBE related PL bands in the range I enlarged mainly, in comparison with the defect related (II) PL bands (Fig.4). At the same time the PL peak position of defect related PL bands shifts into the high energy (to 2.5 eV). With increasing the USP durations the intensity of defect related PL bands (II) raises mainly in comparison with those of NBE PL bands (Fig.4).

A great variety of luminescence bands in the UV and visible spectral ranges have been detected early in the ZnO crystals [18]. The origin of these emissions has not been conclusively established. The NBE emission at 3.0-3.37 eV is attributed to the free (FE) or bound (BE) excitons, to their LO phonon replicas, such as FE-1LO or FE-2LO, or to optical transitions between the free to bound states, such as the shallow donor and valence band, or to donor-acceptor pairs [19].

The blue PL band with the peak at 2.75-2.80 eV is attributed to Cu related defects [23], to Zn interstitials [24] or to donor-acceptor pairs including the shallow donor and oxygen vacancy [24]. The defect related green PL band in the spectral range 2.20-2.50 eV in ZnO is assigned ordinary to oxygen vacancies [24], Cu impurities [23] or surface defects [26]. The orange PL band with the peak at 2.02-2.10 eV was attributed earlier to oxygen interstitial atoms (2.02 eV) [27] or to the hydroxyl group (2.10eV) [28]. Taking into account that the PL intensity of 2.25 eV PL band increased with rising the

USP duration the assumption that the corresponding defects are related to oxygen vacancies looks very reliable.

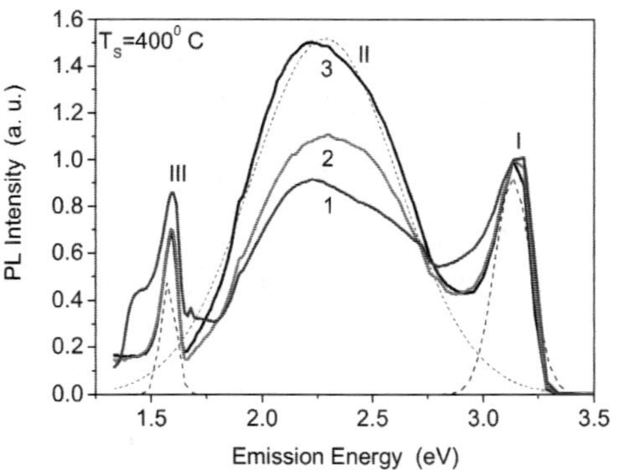

Figure 4a. PL spectra of samples prepared on the substrate with T = 400° C at the deposition times: 1-3 min, 2-5min, 3-10min.

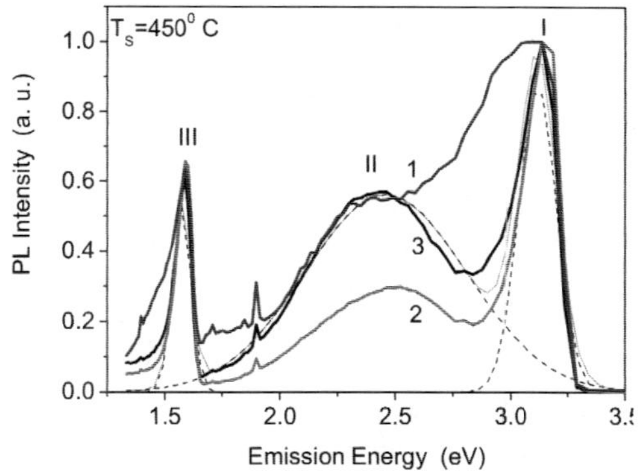

Figure 4b. PL spectra of samples prepared on the substrate with T = 450° C at the deposition times: 1-3 min, 2-5min, 3-10min.

Studied ZnO films were doped by Ag and, therefore, have the acceptor type defects, Ag_{Zn}, which were formed when the Ag atoms substitute of Zn atoms in the ZnO crystal lattice [29, 30]. Emission of the acceptor type defects in ZnO has been studied intensively during the last two decades as well. Note, the broad PL bands with the peaks at 3.238 and 3.315 eV at 4.3K earlier have been observed in acceptor N doped MBE ZnO crystals [31]. These peaks were assigned to acceptor (N_O) bound exciton (3.315 eV), to donor-acceptor pair emission, involving N_O acceptor, or to the LO-phonon replica (3.238eV) of the donor BE line [31]. With temperature increasing the intensity of acceptor bound exciton fallen down due to the dissociation of bound excitons and the FE band or it's LO replicas dominates in the PL spectrum. Thus in our case the 3.14 eV PL band detected at the room temperature can be attributed to the LO phonon replica of FE emission. Note that the variation of PL intensity of the 1.57 eV PL band correlate with the intensity variation of the 3.14 eV PL band that permits to assign of 1.57 eV PL band to the second-order diffraction peak of 3.14 eV PL band.

4. Conclusions

ZnO:Ag NRs with hexagonal structures have been successfully synthesized by the USP method. With increasing the substrate temperature at USP up to 450°C the PL intensity of NBE related emission bands has enlarged. The study has revealed three types of PL bands that in the room temperature PL spectrum related to the LO replicas of FE and its second-order diffraction peak, as well as to the defect-related PL band, apparently, connected with oxygen vacancies. The PL band related to the LO phonon replicas of free exciton in PL spectra measured at room temperature testifies on the high quality of the ZnO:Ag films prepared by USP.

5. References

[1] Pearton S J, Norton D P, Ip K, Heo Y W and Steiner T 2005 *Prog. Mater. Sci.* **50**, 293
[2] Kashiwaba Y, Sugawara K, Haga K, Watanabe H, Zhang B P and Segawa Y 2002 *Thin Solid Films* **411**
[3] Haga K, Suzuki T, Kashiwaba Y, Watanabe H, Zhang B P and Segawa Y 2003 *Thin Solid Films* **433**
[4] Koch M H, Timbrell P Y and Lamb R N 1995 *Semicond. Sci. Technol.* **10** 1523
[5] Vanheusden K, Seager C H, Wareen W L, Tallant D R, Caruso J, Hampden-Smith M J and Kodas T T 1997 *J. Lumin.* **75** 11

[6] Yang Z K, Yu P, Wong G L, Kawasaki M, Ohtomo A, Koinuma H and Segawa Y 1997 *Solid State Commun.* **103** 459
[7] Alvi N H, Usman Ali S M, Hussain S, Nur O and Willander M 2011 *Scripta Materialia* **64**, 697
[8] Huang M H, Mao S and Feick H 2001 *Science* **292** 1897
[9] Scheer R, Walter T, Schock H W, Fearheiley M L and Lewerenz H J 1993 *Appl. Phys. Lett.* **63,** 3294
[10] Chen Y, Baghall D M, Koh H, Park K, Hiraga K, Zhu Z and Yao T 1998 *J. Appl. Phys.* **84,** 3912
[11] Li Y B, Bando Y and Golberg D 2004 *Appl. Phys. Lett.* **84** 3603
[12] Ding J, McAvoy T J, Cavicchi R E, Semancik S 2001 *Sens. Actuat. B* **77** 597
[13] Kaipeng Liu, Beifang F Yang, Hongwei Yan, Zhengping Fu Meiwang Wena,Youjun Chen and Jian Zuo 2008 *Applied Surface Science* **255**
[14] Bengisu Ergin, Elif Ketenci and Ferhunde Atay 2009 *International Journal of Hydrogen Energy* **34**
[15] Kenanakis G, Katsarakis N, Koudoumas E 2013 *Thin Solid Films*
[16] Dybic M, Ostapenko S, Torchynska T V and Velazquez Lozada E 2004 *Appl. Phys. Lett.* **84** (25) 5165
[17] Torchynska T V, Diaz Cano A I, Dybic M, Ostapenko S and Mynbaeva M 2006 *Physica B, Condensed Matter* **376-377** 367
[18] PDF2 XRD database, card no. 36-1451
[19] Djuris A B, Ng A M C and Chen X Y 2010 *Progress in Quantum Electronics* **34** 191-259
[20] Korsunskaya N E, Markevich I V, Torchinskaya T V and Sheinkman M K 1982 *J. Phys. Chem. Solid.* **43** 475-479
[21] Korsunskaya N E, Markevich I V, Torchinskaya T V and Sheinkman M K 1980 *J. Phys. C. Solid St.Phys.* **13** 2975 -2982
[22] Korsunskaya N E, Markevich I V, Torchinskaya T V and Sheinkman M K 1980 *Phys. Stat. Solidi A* **60** 565-572
[23] Reshchikova M A, Morkoc H, Nemeth B, Nause J, Xie J, Hertog B and Osinsky A 2007 *Physica B, Condensed Matter* **401–402** 358–361
[24] Patra M K, Manzoor K, Manoth M, Vadera S P and Kumar N 2008 *J. Lumin.* **128** (2) 267–272
[25] Zhang D H, Xue Z Y and Wang Q P 2002 *J. Phys. D: Appl. Phys.* **35** (21) 2837–2840
[26] Djurišic A B, Choy W C H, Roy V A L, Leung Y H, Kwong C Y, Cheah K W, Gundu Rao T K, Chan W K, Lui H F and Surya C 2004 *Adv. Funct. Mater.* **14** 856-864
[27] Liu X, Wu X, Cao H and Chang R P H 2004 *J. Appl. Phys.* **95** (6) 3141–3147
[28] Qiu J, Li X, He W, Park S-J, Kim H-K, Hwang Y-H, Lee J-H and Kim Y-D 2009 *Nanotechnology* **20** 155603
[29] Chunlei Ren, Beifang Yang, Min Wu, Jiao Xu, Zhengping Fu, Yan lv, Ting Guo, Yongxun Zhao and Changqiong Zhu 2010 *Journal of Hazardous Materials* **182**
[30] Li W J, Kong C Y, Ruan H B, Qin G P, Huang G J, Yang T Y, Liang W W, Zhao Y H, Menga X D, Yua P, Cui Y T and Fang L 2012 *Solid State Communications* **152**
[31] Look D C, Reynolds D C, Litton C W, Jones R L, Eason D B and Cantwell G 2002 *Appl. Phys. Lett.* **81**, 1830

Acknowledgments
The authors would like to thank the CONACYT (project 130387) and SIP-IPN, Mexico, for the financial support, as well as Dr. J. L. Casas Espinola for photoluminescence and the CNMN-IPN for SEM and XRD measurements.

MicroTherm'2013 — Microtechnology and Thermal Problems in Electronics IOP Publishing
Journal of Physics: Conference Series **494** (2014) 012012 doi:10.1088/1742-6596/494/1/012012

Recombination Properties of Diode Structures by Study of Thermal Emission Beyond the Fundamental Absorption Band

T Piotrowski[1], M Węgrzecki[1], A Czerwinski[1], G I Teslenko[2], O Y Malyutenko[2] and V K Malyutenko[2]
[1] Institute of Electron Technology, Warsaw, Poland
[2] Institute of Semiconductors Physics, Kiev, Ukraine

E-mail: piotrows@ite.waw.pl

Abstract. The study presents the possibilities of applying the measurement of spatial and temporal distribution of thermal radiation of a *p-n* junction structure located in a homogeneous temperature field higher than the ambient temperature, modulated by the presence of excess carriers injected through the junction, to determine surface recombination velocity at the injecting contact of the diode emitter and to measure the diffusion length in the base. Good agreement was obtained between the experimental results and calculations based on solutions of the transport equations.

1. Introduction

Performance of *p-n* junction based semiconductor devices (like light modulators and emitters) strongly depends on recombination activity of injection structure elements. We show how mapping the spatial and temporal output of thermal emission (TE) escaping a diode base in the spectral band of free carrier absorption makes it possible to determine such major parameters of the injection process as the effective diffusion length (L) or effective carrier lifetime (τ) in the base and the carrier recombination velocities at the *p-n* junction and ohmic contact planes, s_1 and s_2 respectively.

2. Research methodology

A 5.0 mm-long p^+- n - n^+ germanium structure (figure 1) was taken as a source of infrared thermal emission following free carrier injection and recombination in the structure base.

Figure 1. Image of bulk *n-Ge* based diode with junction on the left and ohmic contact on the right. All free planes are polished and etched. Thermal imaging camera maps the upper plane.

Parameters of current-induced TE output beyond the fundamental absorption band (λ > 2 μm) are related to the base absorption coefficient with the latter being dependent on the concentration of injected free carriers (Δp). TE maps were recorded in continuous (I = 0.5 A, 2D spatial distribution of injected carriers when bias current is on, figure 2) and pulsed (I = 1 A, 200 μs pulse duration) modes with 8-12 μm thermal imaging camera. To fix the temporal evolution of free carriers in the base due to recombination processes, experimental TE maps were recorded at 50, 100, 180, 380 and 580 μs after a bias current is off (figure 3). To have ten times increase of the signal to noise ratio, the study was performed at $T = 75^0$C.

3. Results of measurements and calculations

The 2D distribution of thermal emission output (figure 2) evidences low impact of free carrier recombination at base plane and makes one dimensional approximation valid. Experimental $\Delta p(x,t)$ distribution (figure 3) clearly shows permanent free carrier density decrease due to diffusion and recombination at the ohmic contact for all time delays and still growing impact of this recombination at Δt > 50 μs. The position of maximum concentration region shifts toward ohmic contact when time delay increases. By matching experimental and theoretical results at Δt =100 μs, we obtained following fitting parameters: $L_h = (D_h\tau)^{1/2}$ = 950 μm, s_1 = 180 cm/s, s_2 = 300 cm/s.

Assuming the following simplified relation between injected hole concentration and distance x from the p-n junction plane:

$$\Delta p(x) = \Delta p_0 \exp(-\frac{x}{L_h}) \qquad (1)$$

where Δp_0 is the hole concentration at the junction plane, we obtain carrier concentration distribution determined by the value of electric current and the diffusion length L_h. Inserting this distribution into the formula for emissivity ε and assuming that the absorption coefficient $\alpha(x) = \sigma\cdot\Delta p(x)$, where σ is the cross-section for holes (its value equal to $1.0\cdot10^{-16}$ cm^2 was assumed, taking into account the spectral range of the camera), we obtain the distribution of thermal radiation intensity along the base of the p-n diode using the following relation:

$$\varepsilon(x) = \frac{(1-R)(1-\exp(-\alpha(x)d)}{1-\mathrm{Re}xp(-\alpha(x)d)} \qquad (2)$$

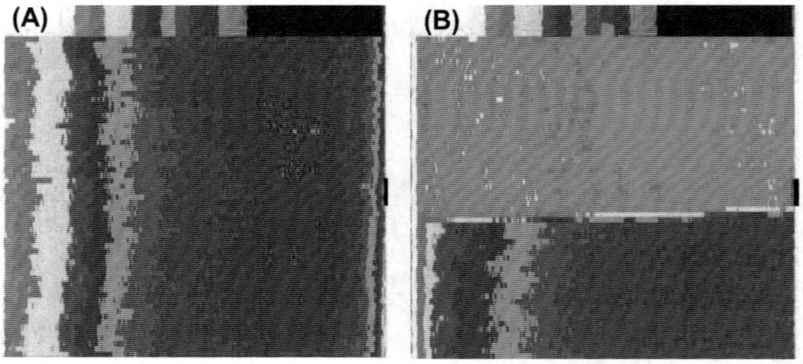

Figure 2. Static thermal emission power map of the diode base (junction is on the left) shown in false colours with red related to higher value. Thermovision images of the structure base, recorded for currents 1A (figure 2A) and 0.5A (figure 2B), at structure temperature 75^{0}C. The upper part of figure 2B shows the uniform temperature distribution on a blackened part of the base. The range of observed changes for thermal radiation intensity in the non blackened part (from the left to the right) correspond to the apparent change of temperature equal to about 10 °C.

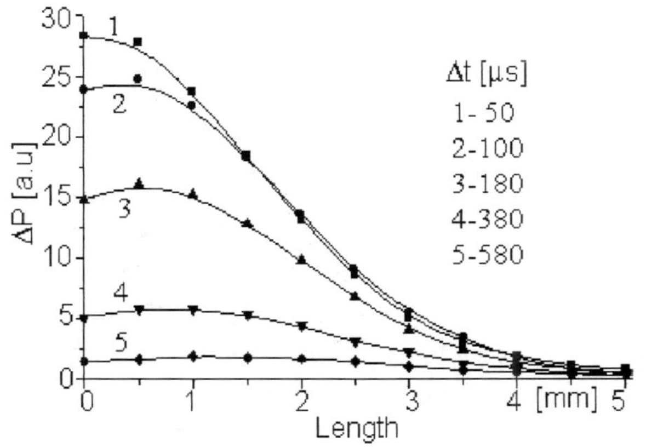

Figure 3. Experimental distribution of thermal emission output along the base (the junction is on the left), measured at several time delays, Δt, after bias current is off.

Taking into account also minor components of the registered radiation (reflection and transmission), depending on the method of measurements, we calculate a theoretical profile of the thermal radiation of the diode base. Figure 4A shows theoretical changes in the distribution of thermal radiation intensity along the base depending on the current value. For low current densities (curve c) this profile is close to the exponential profile, and differs from it at current densities close to the limiting emissivity (1-R) (curves a and b). At high injection levels the entire base of the *p-n* structure reaches the limiting emissivity. Normalizing the theoretical profile P (where $P(x) = W_b \cdot \varepsilon(x)$, with W_b - blackbody total radiant emittance) to unity and comparing it with the experimental profile of the emissivity changes ΔP (curve 1 in figure 3), we can fit both profiles by fixing the parameter L_h. The result of such fitting is shown in figure 4B.

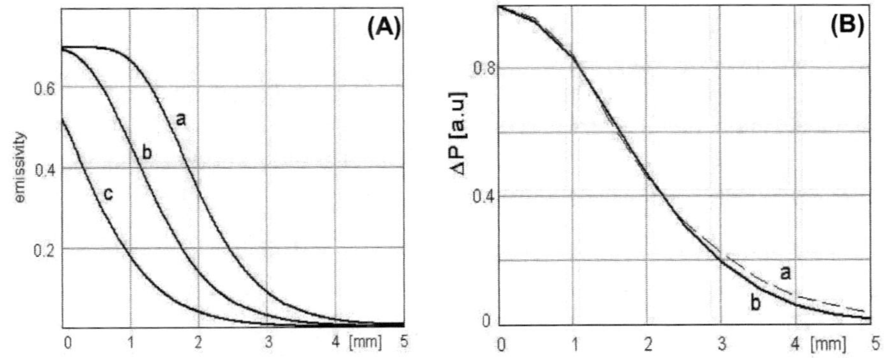

Figure 4. (A) Theoretical distribution of the emissivity along the base from high (a) to low (c) injection levels, (B) Experimental (a) and theoretical (b) profiles of the distribution of thermal radiation intensity obtained with the best fit (i.e. for L_h =950 μm).

In order to calculate the surface recombination velocity at the rectifying contact we will use the measurements of the thermal radiation distribution along the base at 180, 380 and 580 μs time intervals after bias current is turned off, shown in figure 3. These distributions are chosen to allow comparison at low injection levels, where emissivity function is closer to linear.

Spatial distribution of injected holes between the contacts, $\Delta p(x,t)$, and its evolution in space and time were obtained from the solution of the continuity equation:

$$\frac{\partial \Delta p(x,t)}{\partial t} = D_h \frac{\partial^2 \Delta p(x,t)}{\partial x^2} - \frac{\Delta p(x,t)}{\tau} \quad (3)$$

with boundary conditions:

$$D_h \frac{\Delta p}{dx} = s_1 \Delta p \ (x=0), \tag{4}$$

$$D_h \frac{\Delta p}{dx} = -s_2 \Delta p \ (x=w) \tag{5}$$

where D_h is the free hole diffusion coefficient, s_1 and s_2 are the carrier recombination velocities at the junction and ohmic contact planes, respectively, w – the base thickness and τ – the bulk carrier lifetime in the base region. The analyze of the turn-off transient of *p-n* jucction diode was given by R.H. Kingston (see in [3]). The following analytical solution for surface carrier generation, obtained with a multimode method shown in [4], was used for the purposes of the calculations:

$$p(x,t) = 2H \exp\left(\frac{-t}{\tau}\right) \sum_{i=1}^{\infty} \sin\alpha_i \frac{\sin\left(\frac{\lambda_i x}{w} + \alpha_i\right)}{w + D_h\left(\frac{\cos\alpha_i^2}{s_1} + \frac{\cos\beta_i^2}{s_2}\right)} \exp\left(\frac{-(\lambda_i)^2 D_h t}{w^2}\right) \tag{6}$$

with:

$$\lambda = i\pi - \arctan\left(\frac{D_h \lambda}{s_1 w}\right) - \arctan\left(\frac{D_h \lambda}{s_2 w}\right), \tag{7}$$

$$\alpha_i = \arctan\left(\frac{D_h \lambda_i}{s_1 w}\right), \quad \beta_i = \arctan\left(\frac{D_h \lambda_i}{s_2 w}\right) \tag{8}$$

where H – the number of carriers per surface generated directly before the bias current is turned off.

Theoretical distribution of the carrier concentration along the base (the junction is on the left) calculated (in sequence from the top) at 180, 380 and 580 μs time delays after bias current is off, L_h = 950 μm, D_h = 42 cm^2/s is shown in figure 5.

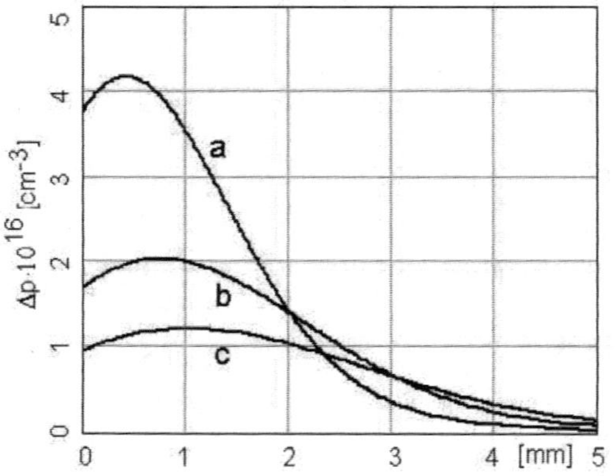

Figure 5. Theoretical distribution of the carrier concentration along the base (the junction is on the left) calculated at (a) 180, (b) 380 and (c) 580 μs time delay after bias current is off. Fitting parameters s_1=180 cm/s, s_2=300 cm/s.

Theoretical analysis shows that clear discrepancies between experimental curves and calculated curves occur at L_h = 950 μm for cases of surface recombination velocity s_1 = 10 cm/s (figure 6A) or for much higher values such as s_1 = 300 cm/s (figure 6B). These significant differences in the shapes of curves indicate the possibility of a fast estimate of contact quality at the injecting junction via observation of the emergence and shift of the thermal radiation maximum at this junction. The further from the injecting contact this maximum occurs, the higher is the recombination velocity.

Since the diode thickness is much larger than the diffusion length, and the surface recombination velocities on polished and passivated surfaces perpendicular to the direction of observation are low, therefore the effective carrier lifetime in the base of the diode is only slightly affected by the surface recombination velocities at these perpendicular surfaces. Therefore the fitted values of carrier lifetime and surface velocities are close to their real values. Even if the surface recombination velocity on these polished and passivated surfaces is as much as 50 cm/s, the reduction of the bulk carrier lifetime is less than 20%.

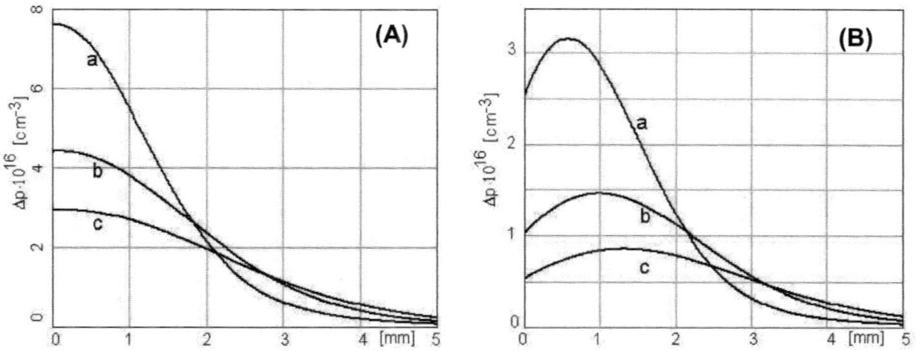

Figure 6. Theoretical distribution of the carrier concentration along the base (the junction is on the left) calculated at (a) 180, (b) 380 and (c) 580 μs time delay after bias current is off. Parameters: s_1=10 cm/s, s_2=300 cm/s for figure 6A, and s_1=300 cm/s, s_2=300 cm/s for figure 6B.

4. Summary

The possibility of measuring surface recombination velocity at contacts and diffusion length in a germanium-based infrared radiation modulator from spatial and temporal distribution of infrared radiation of the heated structure was shown.

This new method enables fast control of contact quality via temporal observation of the spatial position of the maximum infrared radiation intensity relative to the injecting junction plane after forward bias applied to the junction is turned off.

5. References

[1] Piotrowski T *et al* 2011 Optimization of parameters for silicon planar source of modulated infrared radiation, *Materials Science and Engineering B* **176** 363
[2] Malyutenko VK *et al* 2010 Silicon emitter for shortwave infrared (1.6–3 μm) band by light down-conversion, *Appl. Phys. Lett.* **97** 211104
[3] Chih-Tang Sah 1994 *Fundamentals of solid-state electronics* (World Scientific)
[4] Piotrowski T and Sikorski S 1995 Investigation of photoconductivity decay for determination of carrier minority lifetime *Electron Technology* **28** 171

Acknowledgements
This work was supported partially by the Applied Research Programme of the National Centre for Research and Development, project „SiDET"No. PBS1/A9/8/2012.

Journal of Physics: Conference Series **494** (2014) 012013 doi:10.1088/1742-6596/494/1/012013

Optical Fibre Angle Sensor Used in MEMS

J Golebiowski[1], Sz Milcarz[2], M Rybak[2]
[1] Lodz University of Technology, Department of Semiconductor and Optoelectronic Devices, 211/215 Wolczanska , 90-924 Lodz, Poland
[2] Lodz University of Technology, Faculty of Electrical, Electronic, Computer and Control Engineering, 211/215 Wolczanska , 90-924 Lodz, Poland

E-mail: jacekgol@p.lodz.pl, 186238@edu.p.lodz.pl, mrybak@plusnet.pl

Abstract. There is a need for displacement and angle measurements in many movable MEMS structures. The use of fibre optical sensors helps to measure micrometre displacements and small rotation angles. Advantages of this type of transducers are their simple design, high precision of processing, low costs and ability of a non-contact measurement. The study shows an analysis of a fibre-optic intensity sensor used for MEMS movable structure rotation angle measurement. An intensity of the light in the photodetector is basically dependent on a distance between a reflecting surface and a head surface of the fibre transmitting arm, and the deflection angle. Experimental tests were made for PMMA 980/1000 plastic fibres, $\Theta_{NA}=33^0$. The study shows both analytical and practical results. It proves that calculated and experimental characteristics for the analysed transducers are similar.

1. Introduction

There is a need of a displacement and angle measurement in many movable MEMS structures. The use of optical fibre sensors helps to measure micrometre displacements and small rotation angles [1,2,3].

The study shows an analysis of a fibre-optic intensity sensor used for MEMS movable structure rotation angle measurement. Figure 1 shows a basic structure of the transducer. The transducer consists of two arms made of bunches of fibres or a single fibre. An optical source lightens an inspected surface (a mirror surface, e.g. of a silicon cantilever). Reflected light is collected by a receiving fibre and it is transmitted to a photodetector. An intensity of the light in the photodetector is basically dependent on a distance between the reflecting surface and a head surface of the fibre transmitting arm, and the deflection angle.

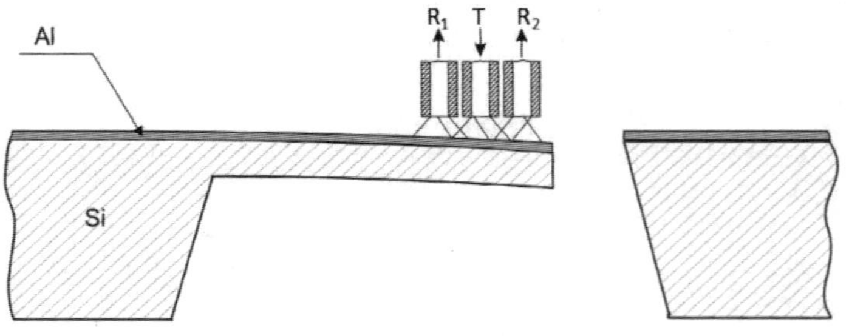

Figure 1. Construction of an optical fibre transducer (where: T – transmitting fibre, R_n – receiving fibres).

2. Analytical model of the sensor

A case of the reflecting surface (reflection coefficient R_m=1) turned through an angle α in the axis of the fibres was analysed in the study (figure 2). For step-index fibres, a Θ_{NA} angle depends on an aperture *NA* of the fibre, a refractive index n_0, and it is described by an equation Θ_{NA} =arcsine (NA/n_0).

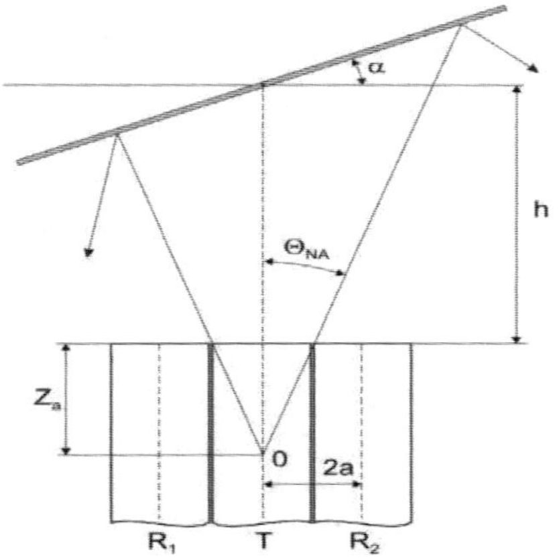

Figure 2. Principle of operation of the fibre optic transducer.

A power P_i coming out from the transmitting fibre is described by a formula $P_i=F_i(n_o)P_t$, where P_t is overall power in the transmitting fibre, $F_i(n_o)$ is the Fresnel's transmission coefficient. The power emitted from the transmitting fibre can be described by the dependence of light intensity *I(r,h)*[4]:

$$P_i = \int_0^R \int_0^{2\pi} I(r,h) r d\phi dr \tag{1}$$

where: r – distance from the axis of the transmitting fibre in a plane perpendicular to the axis, ϕ – azimuth angle, R – radius of the light cone in the distance of $2h$ given by a formula:

$$R=a+2h \cdot tg(\Theta_{NA}) \tag{2}$$

where: a – radius of the fibre.
A distribution of the light intensity *I(r,h)* can be described by the equation [4]:

$$I(r,h) = \frac{2 \cdot P_i}{\pi \cdot R^2(h)} \cdot \left(1 - \frac{r^2}{R^2(h)}\right) \tag{3}$$

An optical power P_o received by the photodetector can be described by the equation [2,5]:

$$P_o(r,h,n_0,\alpha) = 2P_t \int_{R_1}^{R_2} \int_0^{\phi_c} R_m(\lambda,\alpha) \cdot T_i(n_0) T_0(r,h,n_0,\lambda) \cdot \frac{2}{\pi \cdot R^2(h)} \cdot \left(1 - \frac{r^2}{R^2(h)}\right) r d\phi dr \tag{4}$$

where: $\phi_c=arccos\ ((3a^2+r^2)/2a^2)$, R_m – refractive index of the mirror surface.

For the case when the reflecting surface rotates in an axis of the fibres, the light emitted by the transmitting fibres has a shape of a cone. Cross-sections of the arms of the transducer are elliptical. The light intensity of the cone cross-section is described by the equation:

$$I(h,\alpha) = \frac{P_i}{\pi \cdot R(h) \cdot R'(h,\alpha)} \tag{5}$$

where P_i is a power on the transmitting arm output, radius $R(h)$ is described by the equation (2) and $R'(h,\alpha)$ denotes a radius of the ellipse described by the equation:

$$R'(h,\alpha)=\frac{h\cdot\sin\Theta_{NA}}{2}\cdot\left\{\frac{1}{\cos(\alpha-\Theta_{NA})}+\frac{1}{\cos(\alpha+\Theta_{NA})}\right\} \tag{6}$$

For the analysed case the mirror was placed in long-range field ($h>z_a$). Optical power for the receiving fibre's surface, for $h>z_a$, equals:

$$P_o(h,\alpha)=\frac{P_i\cdot a^2\cdot R_m}{R(h)\cdot R'(h,\alpha)} \tag{7}$$

3. Analytical and practical results

To calculate an output signal for a multi-fibre intensity transducer signal, the superposition method was used. The intensity of the light for the branch is a sum of intensities for every single fibre.

Figure 3 shows a dependency of the received optical power as a function of the mirror rotation angle for the step-index fibres – one transmitting and one receiving ($\Theta_{NA}=33^0$, a = 0,49 mm) for the power P_t = 170 μW, R_m = 0,91 for aluminium mirror and the distance h = 1 mm. The distance between the axes of the transmitting and receiving fibres is 1 mm. Figure 4 shows a characteristics for two receiving fibres. The characteristic for one receiving fibre has only a rising slope (figure 3) in difference to the other constructions.

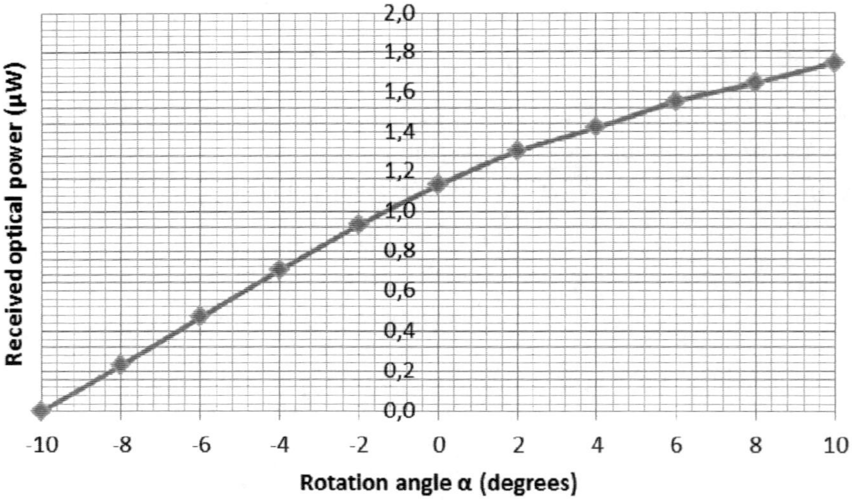

Figure 3. Calculated characteristic of received optical power as a function of a rotation angle, for one receiving fibre.

A practical research was made for PMMA 980/1000 plastic fibres, $\Theta_{NA}=33^0$, with the aid of a test--stand containing a light source (AFL OLS1 – LED, -10dBm output optical power into a 980μm fibre, wave-length 660nm) and a rotating stand equipped with glass mirror and an optical power meter Thorlabs PM 100D with a Si photodiode sensor (400-1100nm).

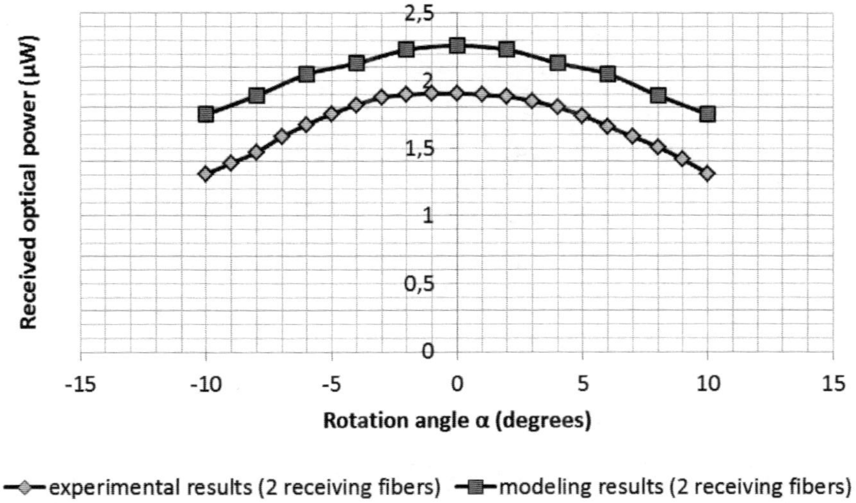

Figure 4. Experimental and modelling results of received optical power versus rotation angle, for two receiving fibres.

In further research an optical head with one central transmitting fibre and four receiving fibres was used (see figure 5). As in the previous case, PMMA fibres were used. The measuring-head was positioned in the distance $h = 1$ mm from the mirror.

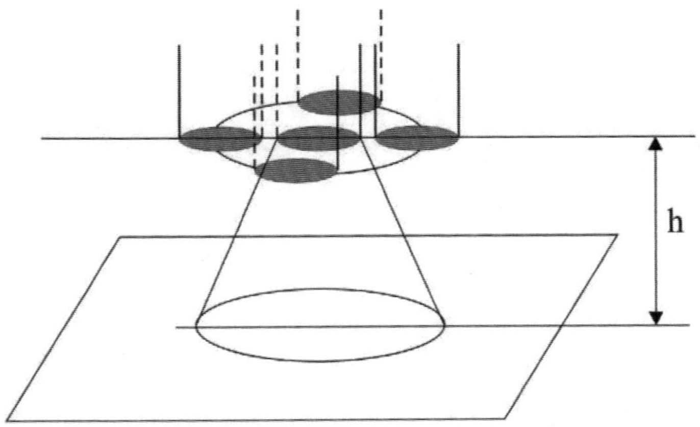

Figure 5. Construction of the optical transducer consisting of one central transmitting fibre and four receiving fibres.

Figure 6 shows analytical results for optical head containing one central transmitting fibre and four receiving fibres (as shown in figure 5). Collected optical power was over 200% higher than for the construction with two receiving fibres. It reached 7.5 μW, while for the two receiving fibres about 2 μW of optical power was observed. The construction was analysed for rotation angles from -10° to +10°. The sensitiveness of this transducer's construction increased in comparison to the construction with two receiving fibres.

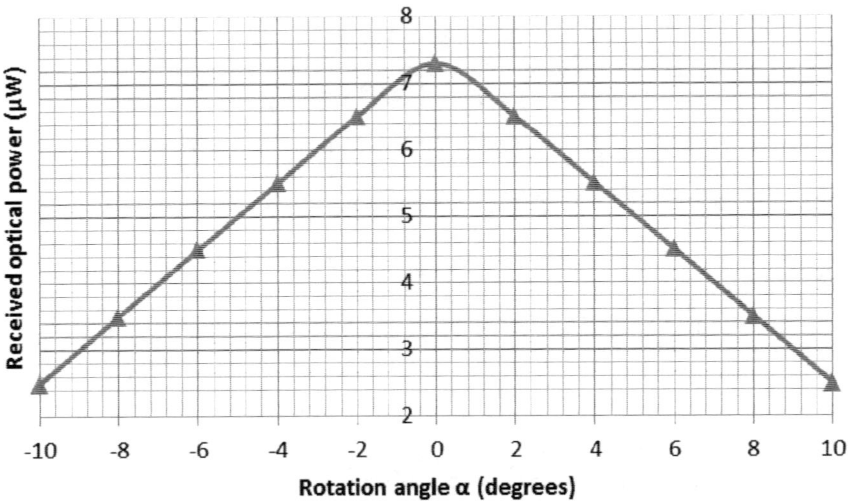

Figure 6. Calculated characteristics for the construction of the transducers containing four receiving fibres.

4. Conclusion

The study shows the analysis of the experimental and numerical results covering different structures of the measuring head (the number of receiving fibres and their placement) and different parameters of the mirrors. The study proves that calculated and measured characteristics for the analysed transducers are similar. Constructions were tested in the range of the rotation angles from -10° to 10°. Received optical powers for the real constructions are slightly lower due to attenuation not considered in the equations used for the calculations. The research shows that maximum optical power for the transducer containing four receiving fibres is over 200% higher than for the construction with two receiving fibres. The surface of the receiving fibres is bigger, so it collects more optical signal. The maximum optical power reached 7.5 μW. The sensitiveness of the transducer also increases in the case of four receiving fibres.

Advantages of this type of the transducers are their simple design, high precision of processing, low costs and ability of a non-contact measurement.

5. References

[1] Wu L, Cheng T, Zhang Q C 2012 A bi-material microcantilever temperature sensor based on optical readout *Measurement* **45** p 1801-1806

[2] Golebiowski J 2012 Microelectromechanical transducer with optical readout for magnetic flux density measurements, *Electrical Review* **no 5** p 22-26

[3] Golebiowski J , Prohun T 2005 *Proceedings of INCINCO Second International Conference on Informatics in Control, Automation and Robotics* (Spain: Barcelona) p 301-306

[4] Suhadolnik A, Babnik A, Mozina J 1995 Optical fiber reflaction refractometer, *Sensors and Actuators B* **29** p 425-432

[5] Binu S, Mahadevan Pillai V.P , Chandrasekaran N 2007 Fibre optic target reflectivity sensor *Opt Quant Electron* **39** p 747-752

Journal of Physics: Conference Series **494** (2014) 012014 doi:10.1088/1742-6596/494/1/012014

Thick Film Heater for Sensor Application

J Milewski[1], M Borecki[1], J Kalenik[1] and K Król[1,2]

[1]Institute of Microelectronics and Optoelectronics, Warsaw University of Technology, Koszykowa 75, 00-655 Warsaw, Poland
[2]Tele- and Radio Research Institute, Ratuszowa 11, 03-450, Warsaw, Poland

E-mail: J.Kalenik@elka.pw.edu.pl

Abstract. A thick film microheater was elaborated. The microheater is intended for fast heating of small volume samples under measurement in optical based system. Thermal analysis of microheater was carried out using finite element method (FEM) for heat transfer calculation as a function of time and space. A nodal heat transfer function was calculated in classical form including all basics mechanisms of heat exchange - heat conduction, convection and radiation were considered. Work focuses on the influence of some construction parameters (ex. length, thermal conductivity of substrate, substrate thickness) on microheater performance. The results show that application of thin substrate of low thermal conductivity and low thickness for miroheater construction and resistor of optimum dimensions leads to significant power consumption decrease and increase of overall optical measurement system performance.

1. Introduction

Optical sensing is one of the most emerging sensor technologies in recent decades. Increasing demand for various data collection is aligned with variety of parameters capable of being measured with the use of optical techniques. The unique optical based system for evaluation of organic liquids using optical capillaries as a sensing element has been developed at Institute of Micro- and Optoelectronics, Warsaw University of Technology. The system was successfully used for investigation of dairy products ex. milk [1] and petrochemical liquids ex. fuel [2]. The system is shown in figure 1. It consists of steel base, laser light source, short optical path part of which is optical capillary filled with investigated liquid, microheater, refracted, transmitted and scattered light detectors. The capillary is placed in shallow V graves on the surface of the steel base. The capillary is pressed to the steel base by two small pieces of magnetic rubber (shown as purple rectangles in figure 1). The heater is placed over the rectangle opening in the steel base (white rectangle in figure 1) on two steps below the surface of the base. The heater is attached to the base by two springs (not shown in figure 1). The dimensions of the heater substrate are 30 x 5 x 0.635 mm. Width of the resistor is 5 mm and length varies in the range from 1 mm to 10 mm. Thickness of the resistor is 20 μm. The level of the steps, the shape of the V graves and the diameter of the optical capillary ensure distance of 50 μm between the surface of the heater and the side wall of the capillary. The liquid placed in quartz capillary is heated over its boiling point by the thick film resistive microheater placed nearby and the bubble of the evaporated liquid is created. The phase change of the medium in capillary, that is a part of the optical path, causes the light transmission change. It is possible to determine many important physical properties of the sample by precise control of temperature of nano-liter volume liquid and transmission of the optical path. For example its purity, chemical composition or percentage of ingredients, can be found. More detailed description of the system was published in [3].

2. Simulation

Since a heat transmission to optical capillary is of great importance in described method the aim of this work was simulation and investigation of thermal properties of heating resistors manufactured using thick-film technology. Simulation was made using finite element method (FEM) for heat transfer calculation as a function of time and space. A nodal heat transfer function was calculated in classical form including all basics mechanisms of heat exchange. No heat transfer through the liquid in capillary was included since the volume of the liquid is negligible in comparison to capillary walls. Heat transfer through the liquid in the capillary is not important because the measurement in the system is carried out until the first bubble of vapour creates and the liquid is removed out of capillary. Al_2O_3 (96%) and ZrO_2 ceramics were used since this substrates have significantly different thermal conductivity with comparable other thermal parameters. First group of heaters was made of standard Al_2O_3 substrates of thickness 635 µm. Repeated measurements of organic liquid properties caused heating of the steel base which had influence on measurements result so the second group of heaters was prepared on ZrO_2 substrates of lower thickness – 150 µm. All important substrate parameters are shown in table 1.

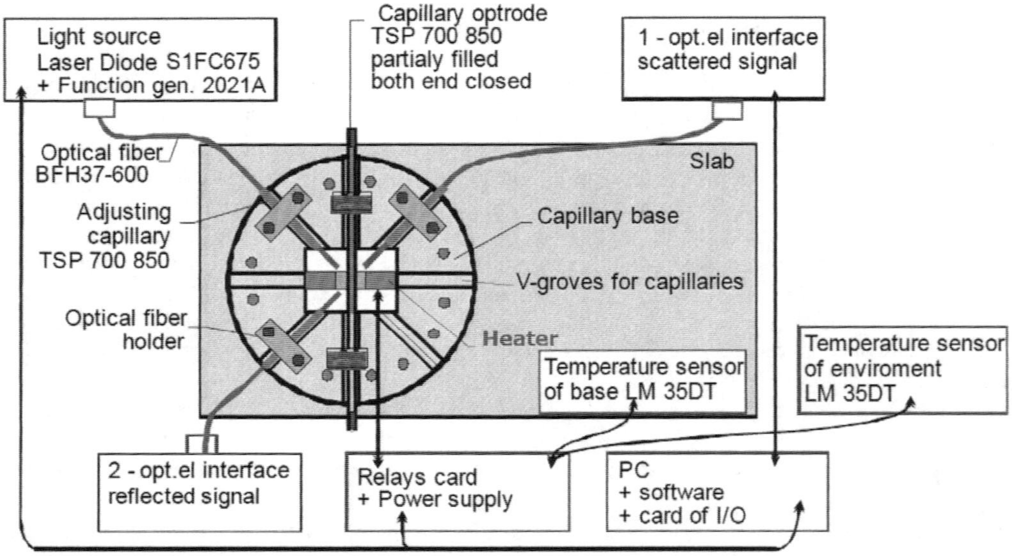

Figure 1. Organic liquid analysis system incorporating optical capillary.

The resistive element was simulated using DuPont 2009 resistive paste with sheet resistance 10 Ω/□ . Thermal conductivity, specific heat and emission coefficient of resistor material are not published by the manufacturer. Thermal conductivity of resistive layer was neglected due to its low thickness. The emission coefficient was assumed 0.9 the same as for ceramic substrate. Calculation focused on influence of resistor length on capillary temperature. Length varied from 1 mm to 10 mm.
The 3D model of microheater was prepared. The model consists of 50 mm diameter and 5 mm thickness steel base with opening in the central part, 30 mm long and 5 mm wide ceramic heater substrate and 150 µm diameter quartz capillary. Thermal conductivity of resistor layer was neglected because of its low thickness. Distance between the thick film resistor surface and the capillary wall was 50 µm. The 3D model of microheater with steel base and optical capillary is presented in figure 2. Simulation of central area of resistor temperature growth was carried out. Applied power was 600 mW/mm^2 i.e. maximum power density for Du Pont 2009 resistive film [4] and power 5W, 4W, 3W, 2W, 1W per resistor. Simulations for maximum power density were carried out to assess

Journal of Physics: Conference Series **494** (2014) 012014 doi:10.1088/1742-6596/494/1/012014

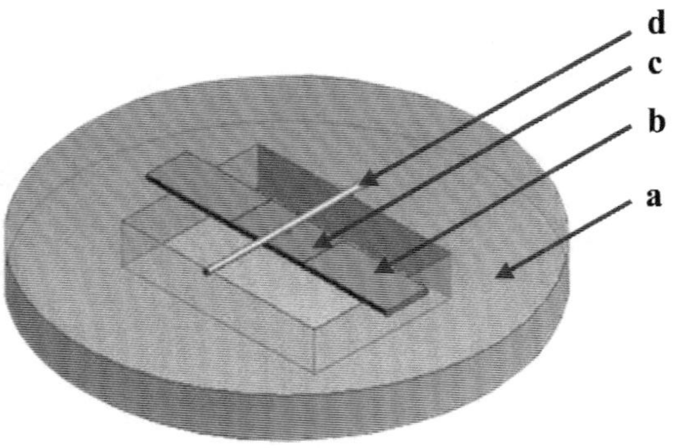

Figure 2. 3D model of microheater: a - stainless steel base, b – microheater substrate, c – thick film microheater resistor, d – optical capillary.

possibility of very fast heating of the object under measurement. Thermal destruction of materials was neglected during simulations. During normal operation power is lower and heating is stopped before the resistive layer reaches temperature of 350 °C. Steel base was the heat sink. Temperature of the steel base was 20 °C, and it was constant during the single experiment as time of heating was very short and amount of heat dissipated to the base did not cause growth of steel base temperature. The contact area between the substrate and the base was 10 mm^2 at each end of the substrate. Convection was natural. The main aim of numerical analysis was reduction of the number of the experiments during resistor dimensions optimization. Figure 3 shows the heater central area temperature dependence for simulations with and without optical capillary for both substrates as a function of resistive element length.

It can be seen that lower heat transfer from the resistor through the substrate to the steel base can cause quicker heating to higher temperature. Lower heat transfer is caused by the use the substrate of lower thickness and lower thermal conductivity.

Presence of capillary in short distance over the resistor reduces temperature growth of central part of the heating resistor.

Table 1. Substrate parameters.

Parameter	Al_2O_3	ZrO_2
Density [g/cm^3]	3.72	6.0
Max. working temperature [K]	1700	1500
Th. conductivity [W/mK]	25	2
Th. Expansion coeff. [10^{-6} K^{-1}]	8.2	10.3
Specific heat [J·kg^{-1}K^{-1}]	880	400
Emissivity	0.9	0.91
Thickness [μm]	635	150

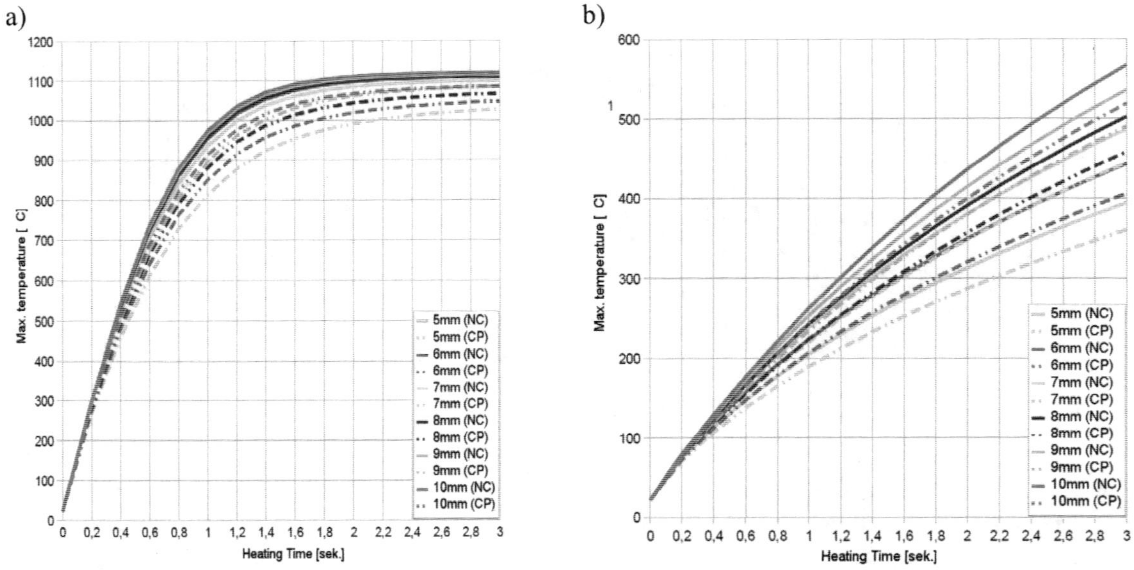

Figure 3. Maximum temperature of heating element as a function of time for ZrO_2 substrate (a) and Al_2O_3 substrate (b). Length of the resistor is the parameter. Solid lines – without optical capillary (NC), dashed lines – with optical capillary (CP).

3. Experimental

For the experiment the microheaters of various resistor length were prepared on alumina and zirconia substrates. Both experiments of heating microheater with and without capillary presence were carried out. Various values of heating power were applied. Temperature of the resistor and capillary surfaces was measured with thermovision camera R300.

The results of temperature distribution measurements over microheater surface with and without capillary presence are presented in figure 4. Microheater was prepared on alumina substrate. Applied power was 5W. Such level of power causes heating the microheater central area up to temperature of 300 °C during about 30 seconds. Quicker heating was possible however the optical measurement setup needed moderate temperature growth. The resistor length was 5 mm. Such resistor length on alumina substrate ensures reasonable balance of radiated power and power dissipated by the substrate to the steel base. Results of simulations and results of measurements are consistent. It can be seen that large amount of power is dissipated through the alumina substrate to the base (see temperature distribution in figure 4c and 4d). Please note higher temperature region of the side surface of the substrate in neighborhood of the resistor (figure 4c). The reason of higher temperature of the side surface of the substrate is low heat transfer through surrounding air in direction perpendicular to the side surface. The two-wedge shape of this region is caused by combination of two heat transfer mechanisms: conduction along the substrate and radiation perpendicular to the substrate surface. As a matter of fact this high temperature region shows the temperature distribution in a cross section of the substrate to some extent. Application of zirconia substrates of low thickness is reasonable. Both simulations and measurements show that capillary presence reduces temperature of the central part of the resistor in proximity of the capillary. It is hard to see, especially at thermal image of the device, because the capillary covers the resistor.

Journal of Physics: Conference Series **494** (2014) 012014 doi:10.1088/1742-6596/494/1/012014

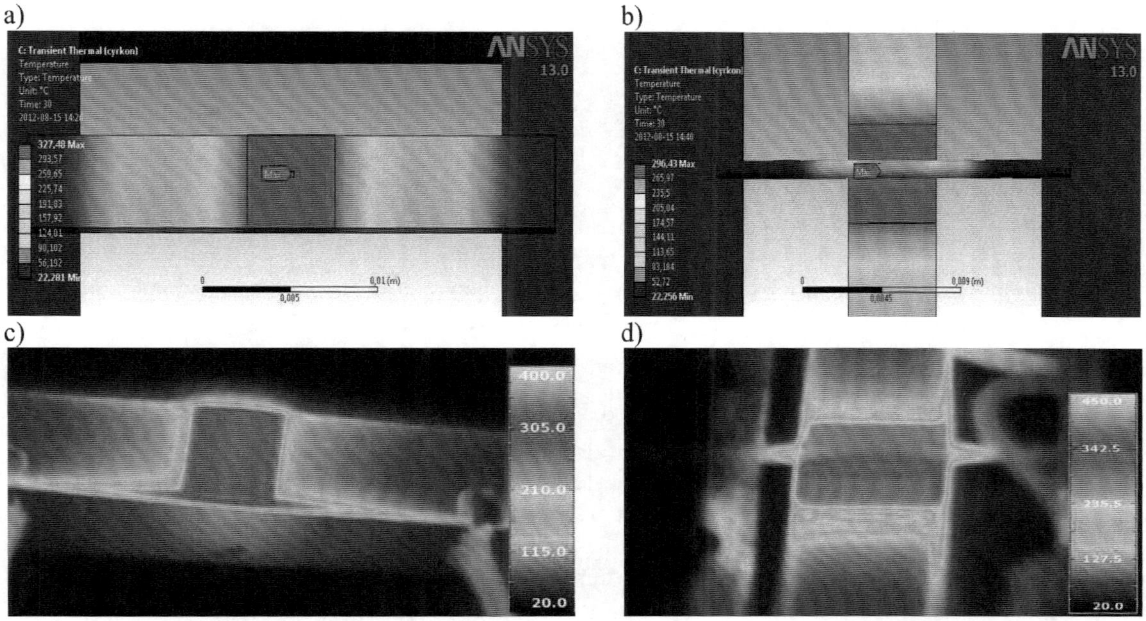

Figure 4. Temperature distribution over microheater surface. Substrate – alumina, resistor length – 5 mm, power - 5 W, time - 30s.: (a) simulation without optical capillary, (b) simulation with optical capillary, (c) thermovision camera image without optical capillary, (d) thermovision camera image with optical capillary.

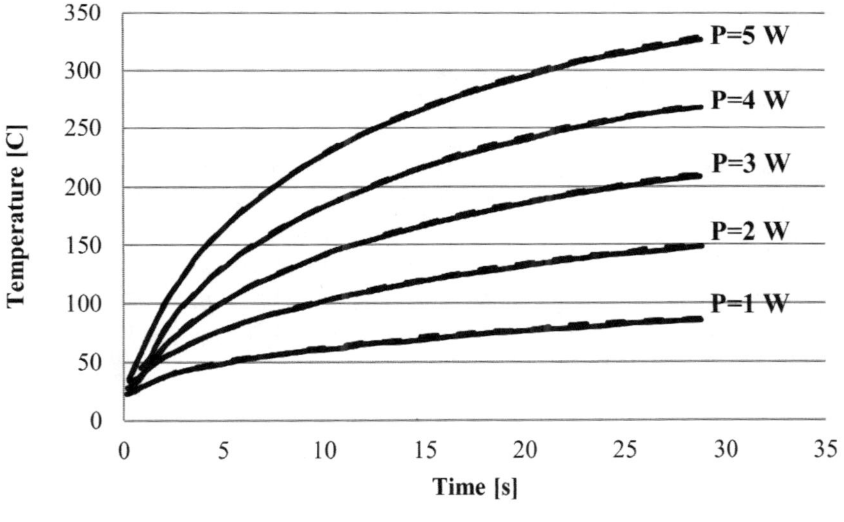

Figure 5. Temperature growth of microheater central area surface. Substrate – alumina, resistor length – 5 mm, power - 5 W. Solid lines – simulation, dashed lines – measurement.

Results of microheater central area temperature growth simulations and measurements for various heating power values are presented in figure 5. Simulations and measurements results are consistent. The good agreement is caused by proper choice of heat transfer mechanisms (conduction through substrate to the steel base, radiation and convection) and parameters for simulation.

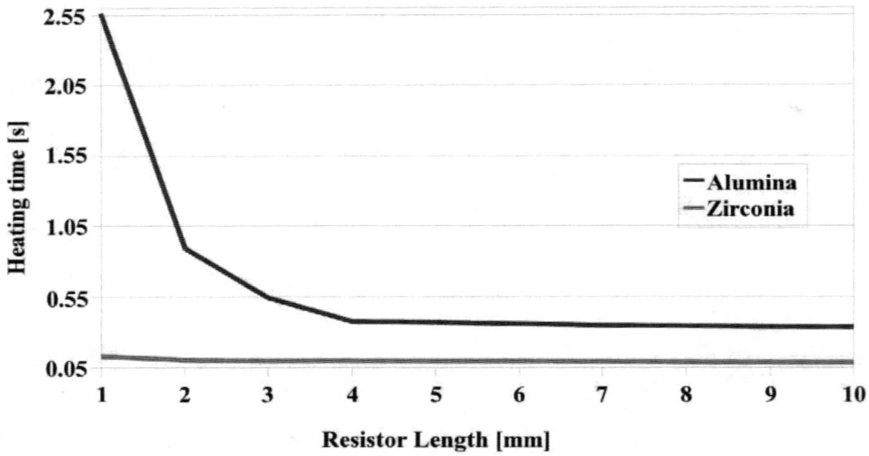

Figure 6. Time needed for heating a water sample to 100°C temperature for alumina and zirconia microheaters as a function of heater length.

Figure 6 presents measured time of reaching temperature of 100°C by the water in the capillary as a function of the resistor length.. The experiment was carried out in optical system. Power density applied to the resistor was 600 mW/mm^2 i.e. maximum power density for Du Pont 2009 resistive film. The moment of reaching boiling point was determined by optical measurement. It can be seen that time of heating to T = 100°C is dependent on resistor length. This time increases for resistors shorter than 5 mm for alumina substrate and 2 mm for zirconia substrate. The reason of heating time reduction is cooling of central area of the resistor by heat transfer through ceramic substrate to surrounding air and to the steel base. Optimum lengths of heating resistors are 5 mm and 2 mm for alumina and zirconia substrates respectively. Resistors of such lengths allow for heating the liquid in the capillary quickly and with minimum power expense.

4. Conclusions

The simulations of temperature growth and temperature distribution over surface of thick film resistive microheater for sensor applications were carried out. The test microheaters were prepared on alumina and zirconia substrates and the heating experiments were executed. The results of performed measurements show that temperature growth and temperature distribution are consistent with simulation results. The best performance was reached for 2 mm long resistor on zirconia substrate. Such construction of thick film resistive microheater allowed to quickly heat the measured liquid sample in the capillary and at the lowest power expense.

5. References

[1] Borecki M, Szmidt M, Pawłowski M K, Bebłowska M, Niemiec T, Wrzosek P 2009 *Photonics Letters of Poland* **1** 37-39

[2] Borecki M, Korwin-Pawlowski M 2010 *"Optical capillary sensors for intelligent classification of microfluidic samples", Nanosensors theory and applications in industry, healthcare and defense* (Warsaw: Taylor and Francis Group)

[3] Kociubiński A, Borecki M 2012 *Elektronika – konstrukcje, technologie, zastosowania,* **9** 91-94.

[4] http://www2.dupont.com/MCM/en_US/assets/downloads/prodinfo/2000.pdf

Acknowledgements

This work was financed by The National Center for Research and Development as the grant No. N R02 0008 06/2010, "New Optoelectronic Devices for Smart Classification of Organic and Biological Liquids".

Journal of Physics: Conference Series **494** (2014) 012015 doi:10.1088/1742-6596/494/1/012015

Semiconductor and Ceramic Microstructure Made by Single Mode Fiber Laser

R Pawlak [1,*], M Tomczyk [1], M Walczak [1], P Domagalski [2]

[1] Lodz University of Technology, Institute of Electrical Engineering Systems Stefanowskiego 18/22, 90-924 Lodz, Poland,
[2] Lodz University of Technology, Department of Chemical Engineering Wolczanska 213, 90-924 Lodz, Poland

E-mail: ryszard.pawlak@p.lodz.pl,
*To whom any correspondence should be addressed.

Abstract. In the paper the results of micromachining of 3D microstructures of microsystems made from silicon and alumina ceramic using a single mode fiber laser (1064 nm) are presented. The quality of obtained structures and its smallest dimensions with acceptable maintained quality were examined. The influence of variable parameters of laser processing with changing of mapping scale on geometrical features of structures was identified.

1. Introduction

3D elements of microsystem are most commonly made of silicon, glass, polymers or ceramic. Main technologies in case of Si are acquired from microelectronics anisotropic wet etching or plasma etching. Wet etching, realized as batch processing, provides high quality of the structures, however, it requires masking and its effectiveness is crystallographic orientation dependent. In case of ceramic materials, which are indispensable for microsystems to work in harsh environment, complex moulding processes or micromachining of green ceramic are used. Among other techniques applied to micromachining of microsystems some subtractive and additive laser technologies have been developed: Laser micromachining LM (2,5D – 3D structures)[1,2], laser micro-stereolitohraphy (LMS)[3] and Direct Laser Interference Patterning (DLIP)[4]. All these techniques are applicable to ceramic materials, while LM only to silicon. Generally the attractiveness of the laser beam as a technological tool follows from an immense surface concentration of power (energy) and the possibilities to influence on the material in a very short time. High spatial resolution achieved by focusing the beam to the size of a few micrometers, local character of the energy interaction, and high flexibility allow to achieve good quality of elements - high precision, small tolerances, precise control and dimensions of components - from several hundred nm to 1 mm. Use of the pulsed nano- pico- and more recently the femtosecond laser beam is presently the standard in laser micromachining. The use of Nd:YAG lasers operating at the second (532 nm), third (355nm) or fourth (266nm) harmonic, Ti: sapphire lasers and the most important types of excimer lasers provides the best results. New types of laser technology, primarily fiber lasers, which today have replaced Nd:YAG lasers in many applications, are also dynamically developing. The effectiveness of thermal ablation of semiconductor and ceramic materials using multimode fiber laser was reported earlier [5,6]. In this paper we discuss the problems connected with microfabrication structures of complex 3D shapes with features of micrometer range by ablation using single mode fiber laser.

2. Experimental part

Two-step experimental tests were carried out in order to determine the smallest dimensions of the ceramic and silicon microstructures produced by laser micromachining with an acceptable maintained quality. In the first step, the analysis of the accuracy of mapping of the elaborated test structure developed by a single line of laser ablation was performed. The effects of laser micromachining were confronted with the results of modelling the ablation process [5]. In the second step – the test structures with different scales were prepared with the process parameters defined on the basis of the first phase of experiment.

2.1. Equipment and process parameters

A single mode fiber laser redENERGY G3 SM 20W (SPI) was used which guaranties the quality of the beam $M^2<1.3$. Laser beam was scanned by 2-Axis Scan Head (Xtreme, Nutfield Techn. Inc.) equipped with F-theta lenses 100mm and 160 mm and software SB-1P Waverunner. The experiments were performed in the following ranges of work parameters: power of the beam – (8-20W); pulse duration – (60-200ns); pulse repetition frequency-(30-90kHz); scanning velocity – (200-4600 mm/s). The beam expander 2,6x and optic head have focused laser beam to the spot of diameter of 50 μm.

a) b)

Figure 1. Equipment used for microfabrication structures (a); microfabrication of capillary separator.

Figure 1 presents a fiber laser redENERGY G3 SM 20W (SPI) with Scan Head and a positioning system. Microfabrication of a capillary separator in silicon is shown in figure 1 b). There is visible characteristic lighting of laser plume. Such treatment is accompanied by sound, what indicates ablation character of this treatment. The maximum efficiency value of this ablation has been presented in table 1 with parameters of nanosecond pulse of 1060 nm fiber laser.

Table 1. Parameters of the processes and efficiency [μm/scan] of laser ablation.

material	efficiency of laser ablation [μm/scan]	Parameters of the process		
		t_i duration of the pulse [ns]	f frequency [kHz]	E_i energy of the pulse [μJ]
Si	~100 μm	220 ns	35	560
Al_2O_3	35	120 ns	36	560

Parameters in table 1 are not sufficient to ensure proper mapping of the pattern in a diminished scale. Scan velocity and frequency become more important in micro scale than efficiency of laser ablation.

2.2. Test pattern

Dimensions of the test pattern were adjusted taking into account the resolution of the used laser beam - spot diameter D = 50 μm (figure 2).

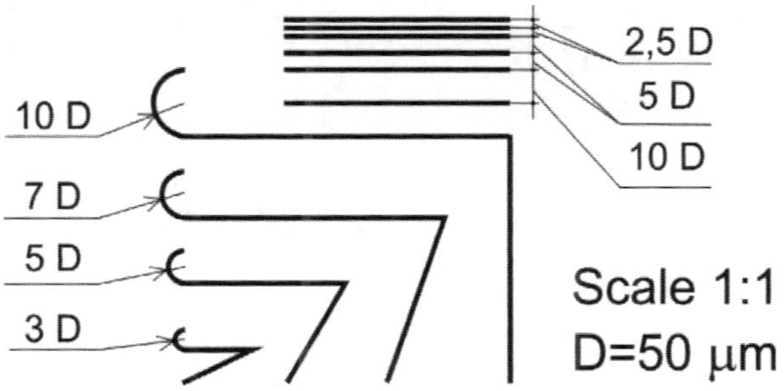

Figure 2. Test pattern.

Features of the test pattern include main problems which occur with laser micro-machined 3D shapes: spatial resolution of adjacent ablation lines affecting hatching for surface treatment; mapping polyline segments with different angles of bending, projection of arcs of small radius of curvature. Dimensions of these "traps" were determined relatively as multiples of the laser beam diameter (figure 2).

2.2.1. Fidelity of the test pattern

Test structures were prepared on silicon and alumina ceramics using conditions of ablative process with moderate dynamics, with a small contribution of the plasma cloud, determined experimentally and confirmed by modelling [5]. For example, an image of a test pattern for a different scale is shown in figure 3 and for different scanning velocity in figure 4.

Variable energetic parameters of the laser beam included average value of power (9-20 W), pulse duration (60 to 200 ns for Si), and repetition frequency (30 - 90 kHz). Re-creation of the test pattern, however, is not a simple reflection of ablative process of material removal, which depends on these parameters. Operation of the laser beam scanner substantially affects mapping quality, which is particularly visible in the manufacture of polylines, and arcs. Images of test structures allow to affirm a beneficial effect of lower scan speeds and optimal power of the beam (12 - 14 W). A clear, local deepening of arcs and sharp bends is observed for the highest power values (16 - 20 W). Application of the smallest power (slightly above the ablation threshold) worsens the continuity of line and uniformity of the depth of the groove, especially for sharp angles.

Figure 3. Pattern mapping for different scale.

Accuracy of the test pattern was evaluated in three-point scale relating to the fulfilment of different criteria: a) mapping of line of test structure - its continuity and uniformity of the depth of the groove, b) mapping of geometric details of shape - sharp angles, arcs, resolution of the pattern. Results of the qualitative assessment of different variables on the process conditions have been summarized in table 2.

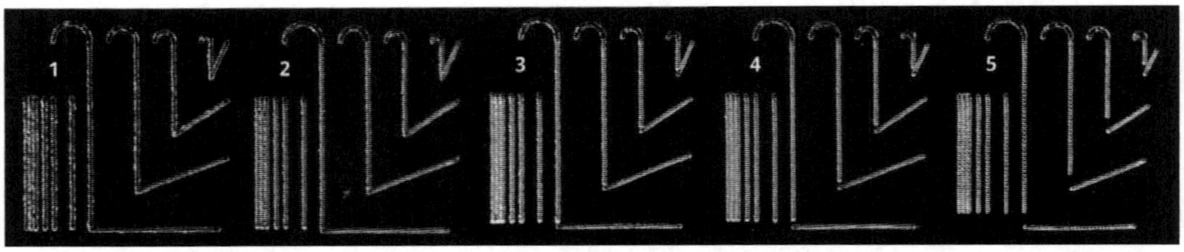

Figure 4. Test pattern for different velocity 1- 500 mm/s, 2- 1000 mm/s, 3- 1500 mm/s, 4- 2000 mm/s, 5- 4000 mm/s.

Table 2. Qualitative assessment of accuracy of the test pattern.

Criterion	Scale				Scanning velocity [mm/s]					Power of laser beam [W]				
	2:1	1:1	1:2	1:4	500	1000	1500	2000	4000	20	14	12	10	9
the continuity of the line	H	H	H	M	H	H	M	M	L	H	H	H	L	L
uniformity	H	H	M	M	H	H	H	H	M	M	M	H	M	L
sharp angles	H	H	H	M	H	H	H	H	H	H	H	H	M	L
arcs	H	H	M	L	H	H	H	H	H	M	H	H	H	M
resolution of line	H	H	L	L	H	M	M	M	M	M	H	H	H	M
Levels of quality criteria: H – high; M – medium; L - low														

2.3. Selecting of microstructures

Among a large number of different elements of microsystems three of them have been selected for testing, and are characterized below. The two test structures (figure 5) were chosen as typical structures used for microfluidics, yet presenting geometry, complicated enough to show capabilities of micromachining technique: the Staggered Herringbone Mixer [7] and phase separator [8]. Scale length ranges of the presented structures were adjusted for typical microfluidics applications, between (approx.) 500 to 50 microns in transverse direction and channel as well as 50-10 micron of length scale of details [9,10]. Both structures present fabrication challenge for typical rapid fabrication techniques like micro milling due to restricted access to bottom wall in case of a mixer, mechanically fragile, narrow splitting "array" for a splitter and generally complicated layout. Typically such structures are fabricated by lithography and/or hot embossing, however, in opposition to presented here laser ablation technique, such techniques cannot be treated as "rapid".

The herring bone mixer (figure 2a) is a commonly used device for mixing in microfluidics. Typical microfluidic flows are deeply laminar, with viscosity dominating over (often negligible) inertia meaning that the fluid flows in layers and all the flow disturbances are damped by viscosity. In such conditions mixing relies on diffusion between fluid layers which is simply slow, therefore mixing in microfluidics is challenging and a couple of novel ways to overcome this problem has been presented [11]. One of them is to force secondary flow pattern for chaotic advection and laminar flow destabilization, perpendicular to main flow direction [12]. That can be done by specific three-dimensional channel layout or by patterned wall structure, as presented in case of herringbone mixer.

The second presented structure (figure 5b) is less popular (yet), however it deals with very practical problem of phase separation in microfluidics. Microfluidic two-phase flow is given a lot of interest nowadays [13] and several successful applications have been shown in this area. The presented device can be used as a splitter utilizing phases surface tension differences as splitting force. The principle of operation is simple: while the two-phase flow is introduced on the inlet, the two single-phase streams are present at two outlets. The geometry, the length scale of passages between the splitting chambers determine the splitting parameters permitting only one phase to overcome the surface tension and pass through.

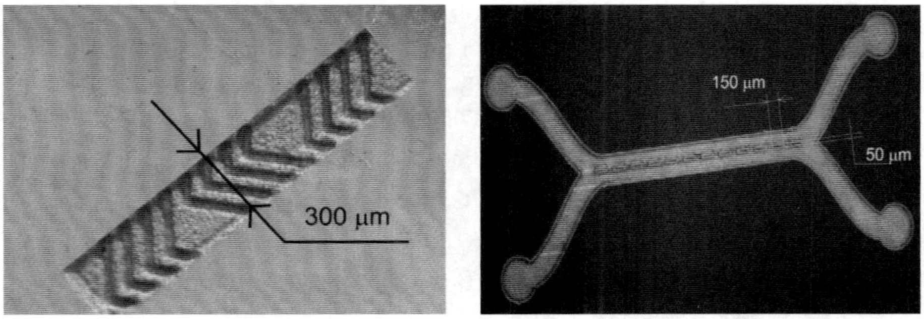

a) **b)**

Figure 5. Structures selected for testing: a - the herring bone mixer; b - capillary separator of two phase mixture.

The third structure (figure 6) is used for micro-channel cooling in semiconducting devices, generating a considerable amount of heat. Dissipation of heat power, higher than 1000W/cm^2 , is possible in such circuits. The greatest cooling efficiency is obtained in two-phase systems with forced circulation of a cooling refrigerant [14, 15]. Such systems with the phase change heat are more efficient than single-phase systems. Optimal use of cooling effect of bubble boiling provides an appropriate choice of the shape and dimensions of the micro-channels. Typical cooling channel widths are between 50 μm to 500 μm with a depth of about 100 μm [16].

a) **b)**

Figure 6. Cooling microchannel: a) width of 500 μm, b) - width of 50 μm.

Laser ablating method for producing micro-channels allows to create micro-channels on a semiconductor substrate directly without any additional pre-processes. Creation of channels of any shape or a resolution lower than a few micrometers is possible depending on the laser source and the optical system. Application of our SPI fiber laser with 1060 nm wavelength resulted in micro-forming the structures of microcoolers with different dimensions and shapes of surface caps- with square or circle cross-section (cylindrical ones). In figure 6 there are shown microcoolers with cylindrical caps of 500 μm and 50 μm width. Fiber laser ablation microtreatment has proved possibility of keeping circular and square shapes even for 50 μm size.

2.4. The quality of the variable scale microstructures

The quality of microstructures produced in variable process conditions is in good agreement with observations for mapping the test structure. There are also additional factors affecting the quality associated with the relationship between a spot diameter (50 μm) and hatching (10, 20, 30 μm) when processing partial surfaces of the structures. Though some roughness of the surface after laser ablation is permissible for large scale, in micro scale it can totally destroy functionality of microdevices. The flow of fluid cannot be correct on undulations on the bottom of channel, resulting from unsuitable selection of the scan velocity, frequency and hatching of the laser pulses (visible on figure 7 and 8).

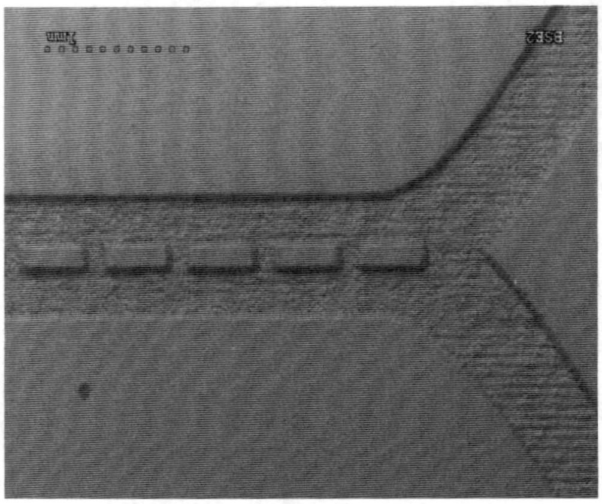

Figure 7. Details of the bottom of capillary separator of two phase mixture - roughness >10 μm.

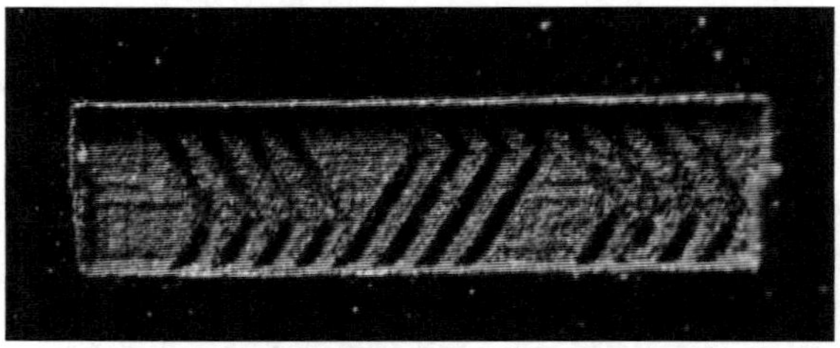

Figure 8. Details of the micro herring bone mixer made by improper selected scan velocity, frequency and hatching-roughness ~10 μm

Figure 9. Micro herring bone mixer made by proper selected scan velocity, frequency and hatching.

Micro herring bone mixer presented in figure 9 proves that proper selection of scan velocity, frequency and hatching of laser beam can help avoid ripples on the bottom of treated region. The sidewalls of grooves are sufficiently smooth for fluid flow, and edges and angles are sharp and precise. Laser ablation has not caused any dangerous mechanical stresses around the treatment zone and there have not been any cracks.

The microstructures of capillary separators of two phase mixture in the scale 1:1. 1:2 and 2:1 have been performed (photos a) and b) in figure 10), using the evaluation results of the quality of mapping the test structures. When the lateral dimensions become closer to the hatching feature, the quality significantly decreases. Small scan speeds contribute to a stronger local heating of the material intensifying the ablation and worsening the quality of the mapped 3D shapes. Increasing energy of pulses with higher scan velocity increases the efficiency of ablation and saves time but there is optimum value of energy for obtaining sufficiently smooth and proper details for effective separation of two phase fluids. Figure 10c presents SEM photos of details made in silicon by an accurate laser beam. There the middle splitting seam is very regular. The bottom of the channel is smooth. To ensure the best quality of the scanned surfaces, required energy should be just on the limit of the ablation threshold for particular materials. For such conditions, during one cycle of laser treatment the removed layer is very thin (even less than 1µm). This allows to determine geometric dimensions accurately and obtain a smooth surface. It should also be remarked that the values of fluence are calculated assuming ideal laser pulse shape in time. According to the manufacturer a real pulse shape is far from perfect rectangle. The given energy value is the mean value throughout the whole duration of the pulse, while its instant values are strongly fluctuating.

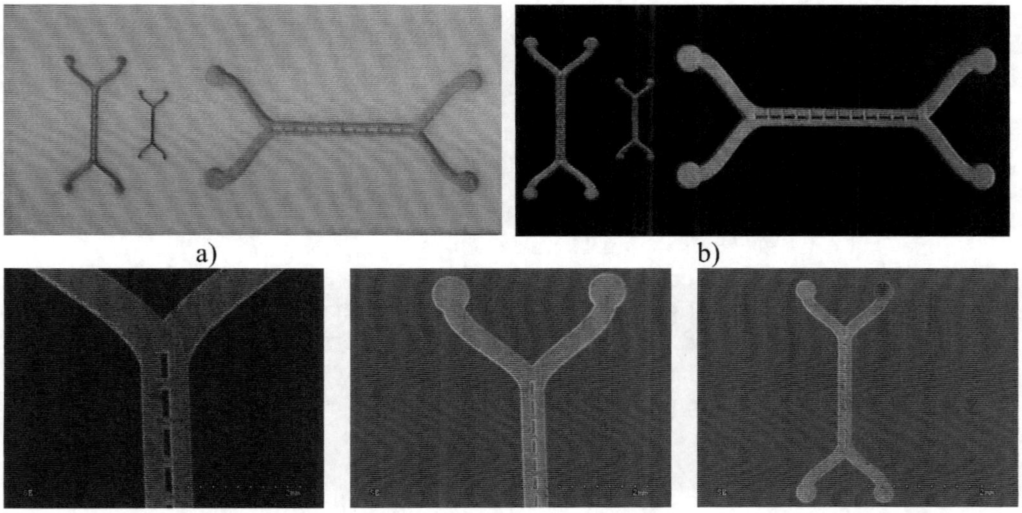

a) b)

c)

Figure 10. Microstructures made by SPI fiber laser ablation in different scale: a – in alumina ceramic; b – in silicon; c – SEM photo of details of smallest structures of (a)

3. Conclusions

The quality of micro-treatment of different microsystems in silicon and alumina was determined by microscopic images (optical and SEM microscopes), whose selection has been presented in figure 4, 5, 6, 9 and 10. In particular the depth of ablation and its uniformity, reproduction of the edges and representation of shapes were assessed. Our experiments have proved that the main obstacles in microtreatment of ceramics and silicon (its brittleness and hardness) can be easily overcome by fiber laser ablation method proposed in section 2.1.

The results of the experiment in manufacturing the microstructures with features of dimensions close to the diameter of nanosecond laser beam, indicate the need to optimize the conditions for such treatment. Optimization must be based not only on the beam energy parameters (reaching the ablation threshold), but also on the properties of the scanner, and the geometric details of the microstructures shapes. Adequate selection of laser beam pulse (duration of the pulse, frequency, scan velocity, hatching and energy) guarantees sufficiently smooth sidewalls and bottoms of the channels as well as sharp edges of pattern lines.

Laser ablation by SPI fiber laser is a rapid process, useful for prototyping and quite convenient because of a relatively low price of such equipment.

4. References

[1] Holthaus M G, Treccani L, Rezwan K 2011 Comparison of micropatterning methods for ceramic surfaces *Journal of the European Ceramic Society* **31** 2809–2817

[2] Samant A N, Dahotre N B 2009 Laser machining of structural ceramics—A review *Journal of the European Ceramic Society* **29** 969–993
[3] Sun C, Zhang X 2002 Experimental and numerical investigations on microstereolithography of ceramics *J. Appl. Phys.* **92** 4796-4802
[4] Marczak J, Rycyk A, Sarzyński A, Strzelec M, Kusiński J, Major R 2013 Direct laser manufacturing of 1D and 2D micro- and submicro-scale periodic structures, *Proc. SPIE* **8703** *Laser Technology 2012: Applications of Lasers* doi: 10.1117/12.2013280
[5] Pawlak R , Rosowski A , Tomczyk M , Walczak M 2013 Laser microscale technologies: Modeling assisted processes *Proc. SPIE* **870** *Laser Technology 2012: Applications of Lasers* art. no. 87030N.
[6] Rosowski A , Sharp M C , Laakso P , Kujanpää V , Pawlak R 2011 The manufacture and processing of ceramic membranes using fiber laser for MEMS system *Proc. of 13th NOLAMP Conference* (Norwegian University of Science and Technology, Trondheim) 293-298
[7] Stroock A D, Dertinger S K W, Ajdari A , Mezic I , Stone H A , Whitesides G M 2002 Chaotic mixer for microchannels, *Science* **295** 647–651
[8] Tokeshi M, Minagawa T, Uchiyama K , Hibara A , Sato K , Hisamoto H and Kitamori T 2002 Single Cell Analysis by Scanning Thermal Lenses *Anal. Chem.* **74** 1565–1571
[9] 2005 *The MEMS Handbook* ed Mohamed Gad-el-Hak (CRC Press; 2nd edition)
[10] Kandlikar, Satish G 2004 Heat transfer mechanisms during flow boiling in microchannels, *Transactions American Society of Mechanical Engineers Journal of Heat Transfer* **126.**1 8-16.
[11] Hessel V , Löwe H , Schönfeld F 2005 Micromixers—a review on passive and active mixing principles *Chemical Engineering Science* **60** 2479 – 2501
[12] Szumbarski J , Blonski S 2011 Destabilization of laminar flow in a rectangular channel by transversely-oriented wall corrugation *Archives of Mechanics* **63**.4 393 - 428
[13] Gunther A , Jensen K F 2006 Multiphase microfluidics: from flow characteristics to chemical and materials synthesis *Lab Chip* **6** 1487–1503
[14] Zhang L, Wang E N, Goodson K E, Kenny T W 2005 Phase change phenomena in silicon microchannels *International Journal of Heat and Mass Transfer* **48** 1572–1582
[15] Gunnasegaran P, Mohammed H A , Shuaib N H , Saidur R 2010 The effect of geometrical parameters on heat transfer characteristics of microchannels heat sink with different shapes *International Communications in Heat and Mass Transfer* **37** 1078–1086
[16] Jung-Yeul Jung, Ho-Young Kwak 2008 Fluid flow and heat transfer in microchannels with rectangular cross section *Heat Mass Transfer* **44** 1041-1049

Acknowledgements
This work was supported by Ministry of Science and Higher Education, Poland, within the framework of Grant N N510 539739 and by National Science Centre Poland, decision no. DEC-2012/05/D/ST8/02389.

Journal of Physics: Conference Series **494** (2014) 012016 doi:10.1088/1742-6596/494/1/012016

Modelling Characteristics of Ferromagnetic Cores with the Influence of Temperature

K Górecki, M Rogalska, J Zarębski and K Detka
Gdynia Maritime University, Department of Marine Electronics, Morska 83, Gdynia, Poland

E-mail: gorecki@am.gdynia.pl

Abstract. The paper is devoted to modelling characteristics of ferromagnetic cores with the use of SPICE software. Some disadvantages of the selected literature models of such cores are discussed. A modified model of ferromagnetic cores taking into account the influence of temperature on the magnetizing characteristics and the core losses is proposed. The form of the elaborated model is presented and discussed. The correctness of this model is verified by comparing the calculated and the measured characteristics of the selected ferromagnetic cores.

1. Introduction

Temperature is one of the most important parameters affecting the properties of electronic components [1 - 4]. In the case of magnetic elements – choking-coils and transformers, the shape of their characteristics is mostly determined by the properties of the ferromagnetic core contained in these components [5 - 8]. As it is known from the literature, e.g. [5, 7, 9] the temperature significantly affects the value of the core saturation flux density and the permeability of this core. When the core temperature exceeds the Curie temperature, the core loses the ferromagnetic properties [5, 10, 11]. It results in a significant decrease in inductance of the choking-coil or in a loss of the voltage feedback between the windings of the transformer.

The models of electronic components play a significant role in the designing process and the analysis of electronic circuits. The accuracy of these models and a set of phenomena taken into account in them determine the validity of the calculations results. Currently, SPICE is the standard program for the analysis of electronic circuits and electric power engineering [6, 12]. Many literature positions concentrate on modelling magnetic elements but they do not include the influence of temperature on the characteristics of the core [12]. Well-known in this area is the classical Jiles – Atherton model which was also implemented in SPICE [10, 13].

As it results from [14, 15], the Jiles–Atherton model has many defects which cause that the modelling magnetization curve does not reflect the character compared with the characteristic of real magnetic cores. Therefore, to describe new ferromagnetic models, the equations from the paper [13] are used. In the model from the paper [13] the Langevin's function is used to describe the magnetizing curve, whereas in the paper [16] the use of the Brillouin's function to this end is proposed.

In the paper [17] the problem of reliable estimation of the values of parameters of the Jiles-Atherton's model, the inaccurate image in this model of more complicated cycles of magnetizing and omission of eddy currents, are pointed out. This problem was described also in other works, for example in the paper [18], where the analysis was performed for the parameters values of the model obtained on the basis of the algorithm proposed in the paper [19]. The authors of the cited papers show that for smaller hysteresis loops, where the difference between the performance of calculations and measurements increases, courses B(t) and H(t) become unsymmetrical and the hysteresis loop does not close. This results from an increase of the magnetic flux density B in the core and a decrease in the magnetic force H, which has no physical justification.

On the other hand, in the literature models of a ferromagnetic core, in the form of a macromodel for SPICE, the thermal properties of cores are taken into account. Such macromodels are shown e.g. in [7, 20 - 23]. Unfortunately, models from the papers [21 - 23] do not include the hysteresis of the core characteristics. Yet the model proposed in [7] contains a lot of errors, which results in the fact that in the core magnetization curve, obtained with the use of this model unjustified high values of the coercive magnetic field and the saturation flux density are observed. The model from [20] correctly describes the influence of temperature on the properties of the ferrite core, but a high complexity of that model prevents its practical application in the analysis of power electronic circuits.

This paper, which is an extended version of the paper [24], proposes a modified model of ferromagnetic core dedicated for SPICE. At the formulation of this model, the classic Jiles-Atherton model from [13], supplemented with the temperature dependence of the parameters of this model taken from [23] and the description of power losses in the core taken from [7] are used.

In the second section the modified ferromagnetic model is presented, whereas the third section contains the results of measurements and calculations of the cores characteristics.

2. Description of the model

Due to the SPICE specification, the elaborated by the authors model is implemented as a subcircuit. The network representation of this model is shown in figure 1. In this model the input signal is put between the terminal H and the ground. This signal represents the magnetic force H. The flux density corresponds to the voltage on the terminal B. The power losses in the core are equal to the voltage at the terminal P_{loss}.

The voltage source E_4 represents magnetization on the initial magnetization curve described by:

$$M_a = \left[\frac{B_{S0}\cdot\left(1+\alpha_{BS}\cdot\left(T_R - T_0\right)\right)}{\mu_0} - H_S\right]\cdot F\left(\frac{H+\alpha\cdot M_a}{A}\right) \tag{1}$$

where B_{S0} denotes the saturation magnetic flux density on the reference temperature T_0 at the magnetic field H_S, α_{BS} – the temperature coefficient of the saturation magnetic flux density changes, T_R – the core temperature, μ_0 – the permeability of free air, α (the magnetic field coefficient) and A (the hysteresis shape parameter). They all are the Jiles–Atherton model parameters. The F(x) function replaces the Langevin function in the classical Jiles–Atherton model and has the following form:

$$F(x) = \frac{x}{\sqrt{x^2+1}}\left[1 - 0.9\cdot\exp\left(-\frac{|x|}{2.5}\right) - 0.1\cdot\exp\left(-\frac{|x|}{25}\right)\right] \tag{2}$$

The current of the voltage source V_{dMa} is equal to the time derivative dM_a/dt.

The voltage source E_5 represents the magnetic flux density expressed by the formula

$$B = \mu_0\cdot\left(H + M\cdot C_R\cdot x\right) \tag{3}$$

where

$$x = \begin{cases} 1 & for\ T_R < T_C \\ 1-\left(T_R - T_C\right)\cdot 0.06 & for\ T_C < T_R < T_C + 17 \\ 0 & for\ T_R > T_C + 17 \end{cases} \tag{4}$$

T_C denotes the Curie temperature.

The current of the voltage source V_{dB} is equal to the derivative dB/dt.

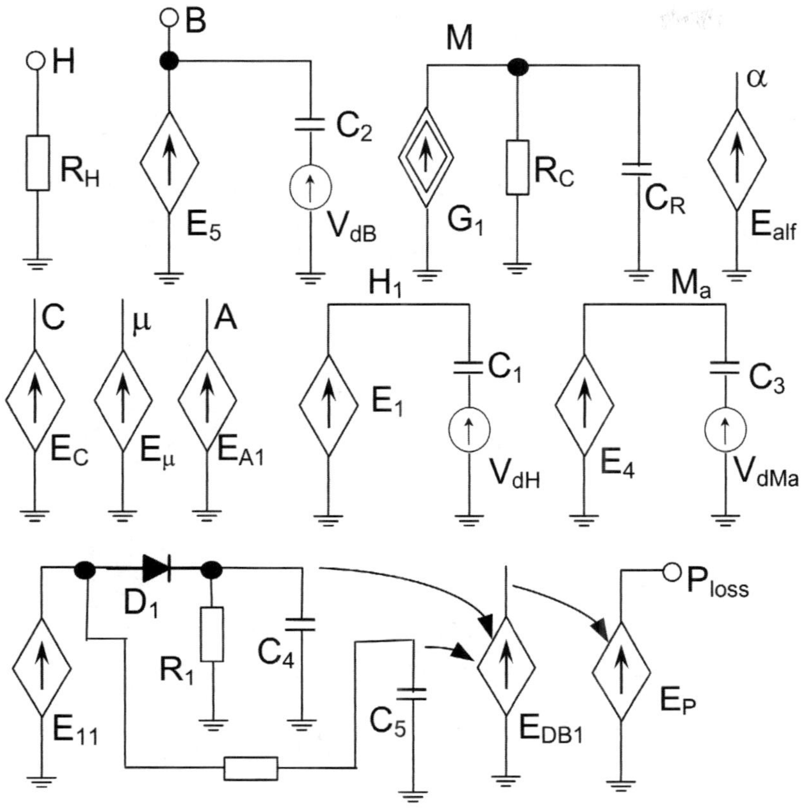

Figure 1. Circuit representation of a new model of ferromagnetic core.

The voltage on the controlled voltage source E_1 is equal to the magnetic field strength H and the current of the voltage source V_{dH} is equal to the derivative dH/dt. The voltage at M node denotes magnetization M in the core. The current on the controlled current source G_1 is described by the formula:

$$G_1 = -\frac{(M_a - M \cdot C_2) \cdot \mathrm{sgn}\left(\frac{dH}{dt}\right)}{C_1 \cdot H_{C0} \cdot [1 + \alpha_{HC} \cdot (T_R - T_0)]} \cdot \frac{dH}{dt} - \frac{C}{C_1 \cdot (1 + C)} \cdot \frac{dM_a}{dt} + M \cdot \frac{C_2}{R_3} \tag{5}$$

where sgn(x) denotes the sign of the x argument, C is the elastic deformation domain wall coefficient, H_{C0} – the coercive field at T_0 temperature and α_{HC} – the temperature coefficient of the coercive field changes. The value of the C parameter is equal to the output voltage of the controlled voltage source E_C described by

$$C = \frac{\mu - 1}{\frac{B_{R0} \cdot (1 + \alpha_{BR} \cdot (T_R - T_0))}{\mu_0 \cdot H_{C0} \cdot [1 + \alpha_{HC} \cdot (T_R - T_0)]} - \mu} \tag{6}$$

where B_{R0} denotes the residual flux density at the reference temperature T_0, α_{RC} – the temperature coefficient of the residual flux density, μ is the core relative permeability described by

$$\mu = \mu_{ip} \cdot [1 + \alpha_\mu \cdot (T_R - T_0)] \cdot x + 1.01 \tag{7}$$

In the dependence (7) - μ_{ip} denotes the initial core permeability at temperature T_0 and α_μ is the temperature coefficient of the initial permeability changes. In turn, the parameter A is equal to the value of the controlled voltage source E_{A1} expressed by the formula

$$A=\left(H_X-\frac{C\cdot M_X}{(1+C)\cdot(\mu-1)}\right)\cdot\frac{1+\alpha_{BS}\cdot(T_R-T_0)}{M_{S0}\cdot(M_{S0}-M_X)-3\frac{M_X}{M_{S0}}} \tag{8}$$

The values of the magnetic field H_X and magnetization M_X appearing in this formula are the coordinates of the point laying on the initial magnetization curve at the reference temperature and M_{S0} corresponding to the saturation value of magnetization for the magnetic field equal to the value H_S at T_0 temperature.

The α parameter is described by the formula

$$\alpha=3\cdot\frac{A}{M_{n0}\cdot[1+\alpha_{BS}\cdot(T_R-T_0)]}-\frac{C}{(\mu-1)\cdot(1+C)} \tag{9}$$

Modelling the core losses P_{loss} by the controlled voltage source E_P is described by the formula from the paper [4] extended with the influence of temperature

$$P_{loss}=\left(\frac{B_{pp}}{2}\right)^{\beta-\alpha}\cdot\left(1+\alpha_P\cdot(T_R-T_m)^2\right)\cdot\frac{P_{V0}}{T}\int_0^T\left|\frac{dB}{dt}\right|^{\alpha}dt \tag{10}$$

In equation (10), α_P denotes the square coefficient of temperature changes of power losses, T_m – the temperature for minimum losses, T is a cycle of induction of the magnetic field and B_{pp} is the peak to peak value of magnetic induction designated in the circuit consisting of the controlled voltage sources E_{11} and E_{DB1},, diode D_{11}, resistors R_{11} and R_{21} and two capacitors C_{11} and C_{21}.

3. Results of calculations and measurements

In order to examine the accuracy of the elaborated model the magnetization characteristics of the selected ferromagnetic cores in the selected temperatures are calculated and compared with the measured characteristics. Additionally, the dependence of the power losses on temperature, amplitude of flux density and frequency were determined. As an example, figure 2 shows the results of calculations (lines) and measurements (points) of the magnetization curve for the ferrite core made of material N27 by EPCOS [9] performed at the frequency f = 10 kHz, whereas figure 3 presents calculated (lines) and catalogue (points) dependences of the core loss on temperature for the same material at the frequency f = 25 kHz. The values of the model parameters, obtained using the algorithm described in [25], are shown in table 1.

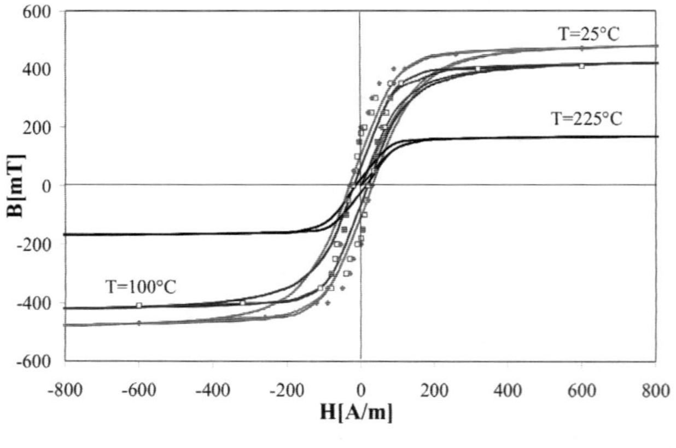

Figure 2. Calculated (lines) and measured (points) magnetization characteristics of the investigated core.

As it is visible, a good agreement between calculated and measured characteristics was obtained. The calculated and measured values of the saturation magnetic flux density, the coercive field and the residual flux density differ each from other no more than 5%. This confirms the correctness of the proposed model. In figure 2, it can be seen that an increase in temperature causes a decrease in the saturation flux density and the narrowing of the hysteresis loop. In turn, figure 3 shows that the core losses are an increasing function of the amplitude of the magnetic flux density. It is worth noticing that for the specified value of the amplitude of the magnetic flux density, the core loses achieve the minimum value at the temperature equal to about 80°C.

Table 1. Values of model parameters for the core made of material N27.

parameter	B_{S0} [T]	α_{BS} [1/K]	H_S [A/m]	T_0 [K]	T_C [K]	H_{C0} [A/m]	α_{HC} [1/K]	P_{V0} [$W/T^{\beta}/s^{-\alpha}$]
value	0.485	-1.65×10^{-3}	1200	300	490	40	-3.33×10^{-3}	1200
parameter	C_1	C_2	B_{R0} [T]	α_{BR} [1/K]	μ_{ip}	α_{μ} [1/K]	M_X [A/m]	T_m [K]
value	10^{-3}	1	0.2	-3.33×10^{-3}	2200	4.67×10^{-3}	2.54×10^{5}	353
parameter	H_X [A/m]	M_{S0} [A/m]	R_3	C_R	β	α	α_P [K^{-2}]	
value	80	3.85×10^{5}	10^{6}	1	2.1	1.214	2.67×10^{-4}	

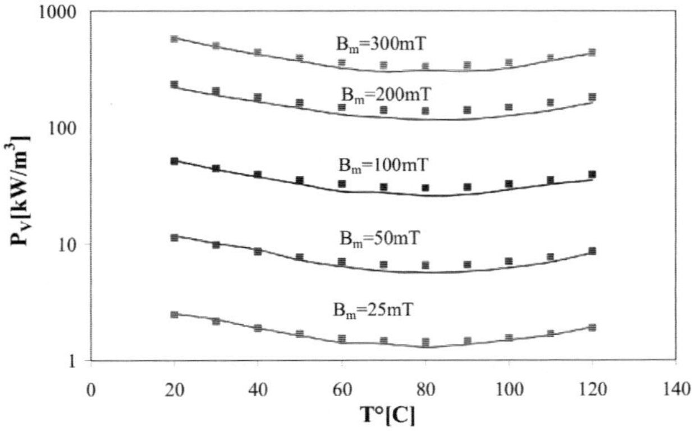

Figure 3. Calculated (lines) and catalogue (points) dependences of a core loss on temperature for material N27.

In figure 4 the calculated and measured dependences of the lossiness of the ferrite core made from material F-867 [26] on the amplitude of the magnitude of the magnetic flux density at the selected values of the temperature and the frequency are shown. The parameters values of the model of this core are collected in table 2.

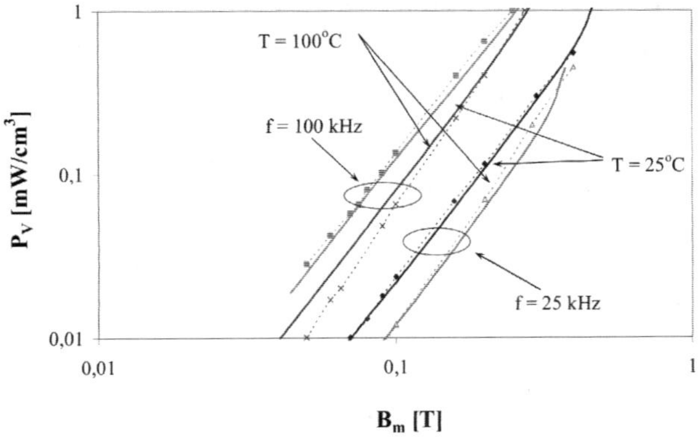

Figure 4. Calculated (lines) and catalogue (points) dependences of a core loss on temperature for material F-867.

As it is visible in figure 4, the good agreement between the results of calculations and measurements is obtained. The lossiness P_V is an increasing function of the amplitude of the magnetic flux density B_m and the frequency f. The dependence $P_V(B_m)$ has a character of the power function with the constant exponent. Only within the range of B_m values close to the values of the saturation magnetic flux density, an increase of the inclination of the considered characteristic is visible. It is worth noticing that in the considered range of changes of B_m and f, the change in temperature value from 25°C to 100°C causes a decrease of the lossiness by even 70%.

Table 2. Values of model parameters for the core made of material F-867.

parameter	B_{S0} [T]	α_{BS} [1/K]	H_S [A/m]	T_0 [K]	T_C [K]	H_{C0} [A/m]	α_{HC} [1/K]	P_{V0} [$W/T^{\beta}/s^{-\alpha}$]
value	0.52	-3.3×10^{-3}	1200	300	513	19.3	-1.3×10^{-3}	2400
parameter	C_1	C_2	B_{R0} [T]	α_{BR} [1/K]	μ_{ip}	α_{μ} [1/K]	M_X [A/m]	T_m [K]
value	10^{-3}	1	0.095	-4×10^{-3}	1630	7×10^{-3}	3.38×10^{5}	361
parameter	H_X [A/m]	M_{S0} [A/m]	R_3	C_R	β	α	α_P [K^{-2}]	
value	200	4.126×10^{5}	10^{6}	1	2.286	1.46	2.2×10^{-4}	

4. Conclusions

In the paper the modified model of the ferromagnetic core is proposed. This model is based on the basic equation described by Jiles and Atherton. The modified model possesses a different analytical description of the initial magnetization curve where the submission of the exponential function instead of the classic Langevin's function is used. In addition, the influence of temperature on the magnetizing characteristics of the core is taken into account. Moreover, this model contains the description of the core losses resulting from the core remagnetization. The presented model uses only 16 parameters, but most of them can be calculated from the core catalogue.

The results of calculations and measurements, presented in the previous sections, confirmed the correctness of the description of the magnetizing characteristics in the wide range of temperature changes and the dependence of the core losses on the amplitude of the flux density, frequency and temperature. The presented results of calculations and measurements refer to the ferrite cores. The authors performed analogical calculations for ferromagnetic cores made of different materials and they obtained the good agreement between calculated and measured characteristics. The calculated and measured characteristics of the investigated ferromagnetic cores have the same shape and the coordinates of the characteristic points laying on these characteristics not significantly differ from each other. The obtained results prove the correctness of the presented model.

5. References

[1] Górecki K and Zarębski J 2012 The method of a fast electrothermal transient analysis of single-inductance dc-dc converters *IEEE Trans. on Power Electronics* **27** (9) 4005-4012

[2] Castellazzi A, Gerstenmaier YC, Kraus R and Wachutka GKM 2006 Reliability analysis and modeling of power MOSFETs in the 42-V-PowerNet *IEEE Trans. on Power Electronics* **21** (3) 603-612

[3] Zarębski J and Górecki K 2009 SPICE-aided modelling of dc characteristics of power bipolar transistors with selfheating taken into account *International Journal of Numerical Modelling Electronic Networks, Devices and Fields* **22** (6) 422-433

[4] Wac-Włodarczyk A 2012 Materiały magnetyczne. Modelowanie i zastosowania *Monografie – Politechnika Lubelska*

[5] Van den Bossche A. and Valchev VC 2005 Inductors and transformers for power electronics *CRC Press, Taylor & Francis Group Boca Raton*

[6] Maksimovic D, Stankovic AM, Thottuvelil VJ and Verghese GC 2001 Modeling and simulation of power electronic converters *Proceedings of the IEEE* **89** (6) 898-912

[7] Wilson PR, Ross JN and Brown AD 2002 Simulation of magnetic component models in electric circuits including dynamic thermal effects *IEEE Trans. on Power Electronics* **17** (1) 55-65

[8] Ericson R and Maksimovic D 2001 Fundamentals of power electronics *Norwell, Kluwer Acad. Publ.*

[9] 2000 SIFERRIT materials *Application notes EPCOS AG*

[10] Miljevec D and Zidaric B 2008 Introducing a domain flexing function in the Jiles – Atherton hysteresis model *Journal of Magnetism and Magnetic Materials* **320** 763 – 768

[11] Pressman AI 1998 Switching power supply design *McGraw-Hill, New York*

[12] BassoC 2001 Switch-mode power supply SPICE cookbook *McGraw-Hill New York*

[13] Jiles DC and Atherton DL 1986 Theory of ferromagnetic hysteresis *Journal of Magnetism and Magnetic Materials* **61** 48-60

[14] Zirka S E, Moroz YI, Harrison RG and Chwastek K 2012 On physicl aspects of the Jiles-Atherton hysteresis models, *Journal of Applied Physics* **112** 043916-1 – 043916-7

[15] Chwastek K 2010 Problems in descriptions of hysteresis *Przegląd Elektrotechniczny* **86** (4) 24-27

[16] Boukhtache S, Azoui B and Féliachi M 2006 A novel model for magnetic hysteresis of silicon-ironsheets *Eur. Phys. J. Appl. Phys.* **34** 201-204

[17] Chwastek K 2008 Frequency behaviour of the modified Jiles – Atherton model *Physica B* **403** 2484-2487

[18] Lederer D, Igarashi H, Kost A and Honma T 1999 On the parametr idenctifocation and application of the JA hysteresis model for numerical modelling of measured characteristic *IEEE Trans. on Magnetics* **35** (3) 1211-1214

[19] Corana A, Marchesi M, Martini C and Ridella S 1995 Minimizing multimodal functions of continuous variable with the 'simulated annealing' algorithm *ACM Trans. on Mathematical Software* **31** 4306-4311

[20] Górecki K 2003 Modelowanie cewki z rdzeniem ferrytowym w programie SPICE z uwzględnieniem samonagrzewania *Kwartalnik Elektroniki i Telekomunikacji* **49** (3) 389-404

[21] Górecki K and Zarębski J 2009 Electrothermal analysis of the self-excited push-pull dc-dc converter *Microelectronics Reliability* **49** (4) 424-430

[22] Górecki K 2008 Elektrotermiczny model cewki z rdzeniem ferrytowym do analizy przetwornic dc-dc *Elektronika* **49** (4) 194-198

[23] Górecki K and Detka K 2012 Electrothermal model of choking-coils for the analysis of dc-dc converters *Materials Sciences & Engineering B* **177** (15) 1248-1253

[24] Górecki K, Detka K, Zarębski J and Rogalska M 2013 Modelling the influence of temperature on the characteristics of the ferromagnetic core *Proceedings of Microtherm 2013 Microtechnology and Thermal Problems in Electronics* 99-105

[25] Górecki K, Rogalska M and Zarębski J 2013 Parameter estimation of the electrothermal model of the ferromagnetic core *Proceedings of the 20th International Conference Mixed Design of Integrated Circuits and Systems MIXDES* 435-439

[26] Power materials characteristics *http://www.aet.com.pl/Portals/0/ZelNet_HermesNet/ProductFiles/CHARAKTERYSTYKI%20MATERIA%C5%81%C3%93W%20MOCOWYCH.pdf*

Acknowledgements

This project is financed from the funds of National Science Centre which were awarded on the basis of the decision number DEC-2011/01/B/ST7/06738.

Journal of Physics: Conference Series **494** (2014) 012017 doi:10.1088/1742-6596/494/1/012017

Design and Modelling of a Silicon Optical MEMS Switch Controlled by Magnetic Field Generated by a Plain Coil

J Golebiowski[1], **Sz Milcarz**[2]
[1] Lodz University of Technology, Dept. of Semiconductor and Optoelectronics Devices, 211/215 Wolczanska, 90-924 Lodz, Poland
[2] Lodz University of Technology, Faculty of Electrical, Electronic, Computer and Control Engineering, 211/215 Wolczanska, 90-924 Lodz, Poland

E-mail: jacekgol@p.lodz.pl, 186238@edu.p.lodz.pl

Abstract. Optical switches can be made as a silicon cantilever with a magnetic layer. Such a structure is placed in a magnetic field of a planar coil. There is a torque deflecting the silicon beam with NiFe layer depending on a flux density of the magnetic field. The study shows an analysis of ferromagnetic layer parameters, beam's dimensions on optical switch characteristics. Different constructions of the beams were simulated for a range of values of magnetic field strength from 100 to 1000 A/m. An influence of the actuators parameters on characteristics was analysed. The loss of stiffness of the beam caused by specific constructions effected in displacements reaching 85 nm. Comsol Multiphysics 4.3b was used for the simulations.

1. Introduction

The use of optical sensors in the industry is still expanding. A transmission of a signal from the sensors is mostly done by optical fibers. Splitters and optical switches are used to minimize the amount of elements and ease the process of collecting data from optical circuits [1,2]. Another advantage is a possibility to transmit measured data from many sensors for different wavelengths by the minimized number of optical paths. The development of the technology effects in smaller sensors and an increased amount of sensors in a single structure. Increasing a number of bistable switches will be necessary [3]. A replacement of a bistable micro-switches matrix by multi-state switches will effect in a loss of power consumption and dimensions of a whole layout, and a simplification of a controlling system. The optical switches are made as a silicon cantilever with a magnetic layer (figures 1, 2). Such a structure is placed in a magnetic field of a planar coil. There is a torque deflecting the silicon beam with NiFe layer depending on a flux density of the magnetic field [4,5].

A position of the switch is controlled by a change in an external magnetic field $\boldsymbol{H}$ (a change in a current in the coil) which effects in a torque $\boldsymbol{T_H}$ deflecting the beam, which can be described by the equation [6]:

$$T_H = \mu_0 V\left|\vec{M} \times \vec{H}\right| = \mu_0 V \cdot M \cdot H \cdot \sin\alpha \qquad (1)$$

Where : $\boldsymbol{M}$ - magnetization vector of a the ferromagnetic layer, α - an angle between the magnetization vector and the external magnetic field vector $\boldsymbol{H}$, V - volume of the magnetic material

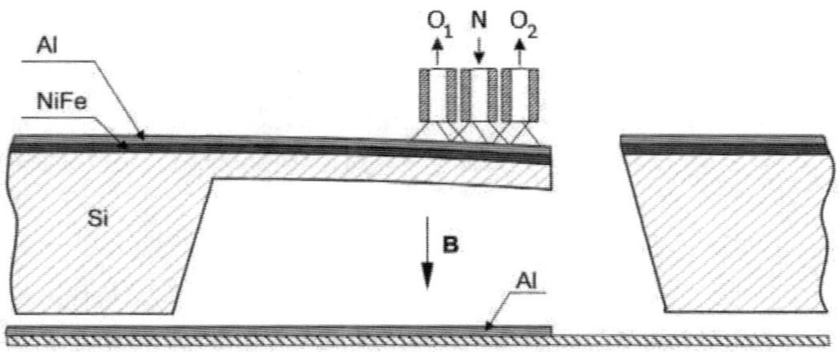

Figure 1. Construction of MEMS optical switch.

2. Modelling and simulation

Due to a complicated construction of MEMS (many layers of different materials), silicon anisotropy and non-linear ferromagnetic characteristics, analytical calculations are too complicated. The Comsol Multhiphysics and the finite elements method were used in the study. In the models, combined mechanical and magnetic fields were used (MEMS, AC/DC modules) [7,8]. A 3D magnetic field distribution generated by the planar winding was considered [9]. With the magnetization vector it effects in the moment bending the beam. The study shows an analysis of numerical models of MEMS transducers used in silicon microactuators and controlled by magnetic field. Movable silicon cantilevers structures with a ferromagnetic layer on the surface, which were attached from one side, were considered. A mirror placed on a free end of the beam reflected transmitted light. Silicon cantilevers (Si <100>) (E=160 GPa, density 2320 kg/m^3, ε_r=4.5), thickness 10 µm, width 2- 3 mm and length 4- 6 mm with 10 µm ferromagnetic layer (E=152 GPa, density 7860 kg/m^3, ε_r=1) were analysed (figure 2). Further reduction of the thickness caused errors in the simulation, especially on a boundary between the silicon and ferromagnetic layer. The models were meshed using very fine free triangular mesh. Mapped meshes were also tested, but they caused serious errors on the boundaries between the used materials. The simulations were computed on Intel i7, 6 core machine with 12 GB of RAM memory. Single simulation for 3 "k" factors and range of magnetic field strength was taking from one to two hours. The k factor represented anisotropic properties of the ferromagnetic layer. In case where length/thickness and width/thickness proportion are much higher than 100 and length of the layer is bigger than its width, geometrical properties of the layer have crucial influence on the k factor. The k factors of values from 0.05 to 0.2 were calculated according to geometrical properties and magnetic properties of NiFe layer.

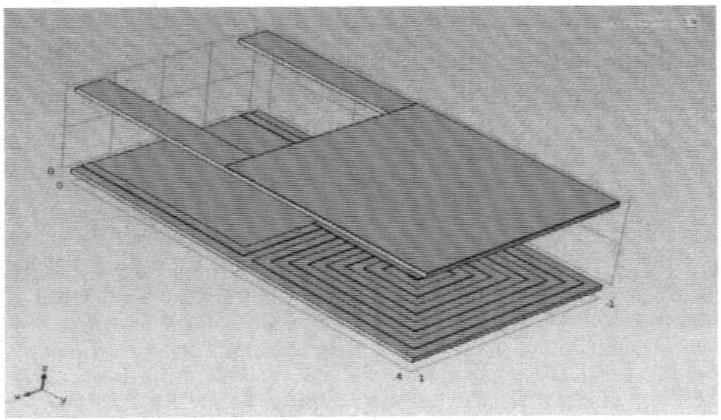

Figure 2. Model of a switch geometry in the Comsol.

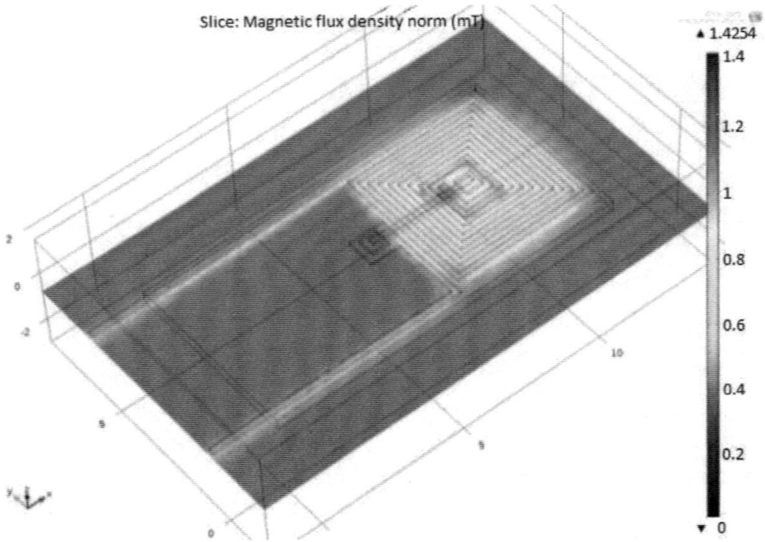

Figure 3. Distribution of magnetic flux density on surface coil (normal axis *Bz*) (DC).

Figure 4 shows deflection of a 10 μm thick, 3 mm wide and 6 mm long beam, with 50% Ni 50% Fe 10 μm layer on its surface, in strong magnetic field *B*=1 mT (field direction is perpendicular to the beam's surface). The beam's deflection reaches almost 85 nm. However, further research was carried out for lower values of magnetic field, which generation in real planar coil–constructions is more probable. The influence of magnetic field on the beam's deflection was further analysed.

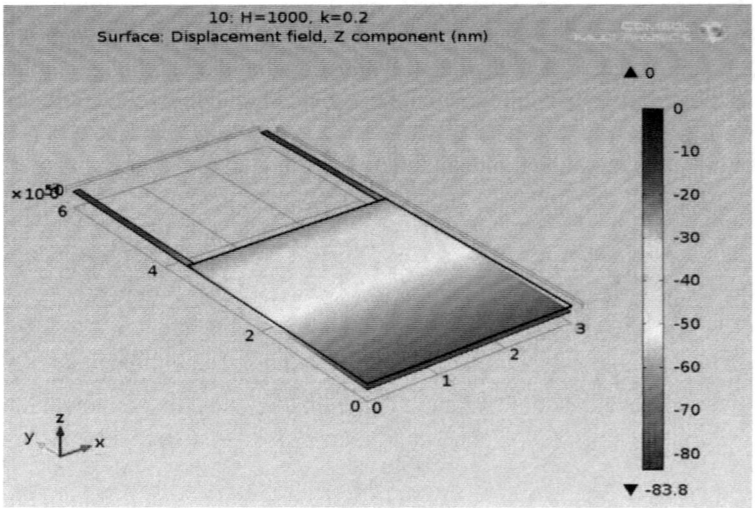

Figure 4. Deflection of the cantilever with a ferromagnetic layer in a magnetic flux density B=1 mT (deflection scale factor 1000:1).

3. Results of modelling and simulation

This paragraph shows the results of modelling of different 3x6 mm 10 μm cantilever constructions: a full cantilever, a cantilever with a 2.6x2.5 mm cut-out cavity and an improved construction of cantilever with 2.8x2.5 mm cut-out cavity. Each of the beams had a 10 um ferromagnetic layer on its surface. The cantilevers were put and simulated in magnetic field varying from 100 to 1000 A/m.

Figure 5 shows results of the modelling for the most stiff construction – 3x6 mm, 10 μm thick full cantilever with 10 μm ferromagnetic layer on its surface. A maximum deflection is about 1 nm for reasonable values of magnetic field and maximum k factor equalled to 0.2. The full beam construction occurs to be too stiff in comparison to the cut-out construction, which use is much more relevant (compare figures 5, 6 and 7).

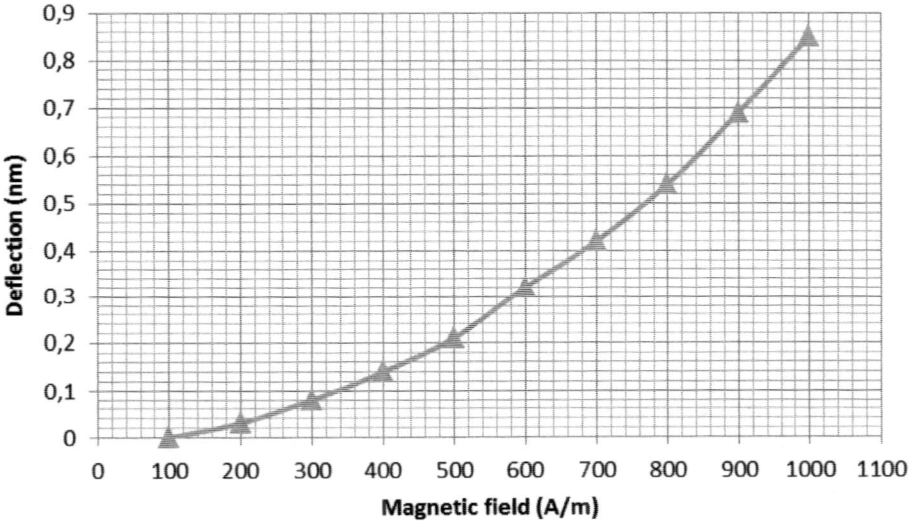

Figure 5. Deflection of the full 3x6 mm cantilever as a function of magnetic field for k=0.2.

Figures 6 and 7 show results of simulation for the cut-out 3x6 mm (10 μm silicon, 10 μm ferromagnetic layer) cantilevers' constructions for different values of magnetic field and k factors. The loss of stiffness effects in much higher deflections, for reasonable magnetic fields, reaching 43 nm for the basic construction (2.6x2.5 mm cut-out cavity) and almost 85 nm for the improved construction. The second mentioned constructions has an enlarged 2.8x2.5 mm cut-out cavity, which means that a surface where ferromagnetic layer is placed is attached by two 0.1x2.5 mm silicon beams.

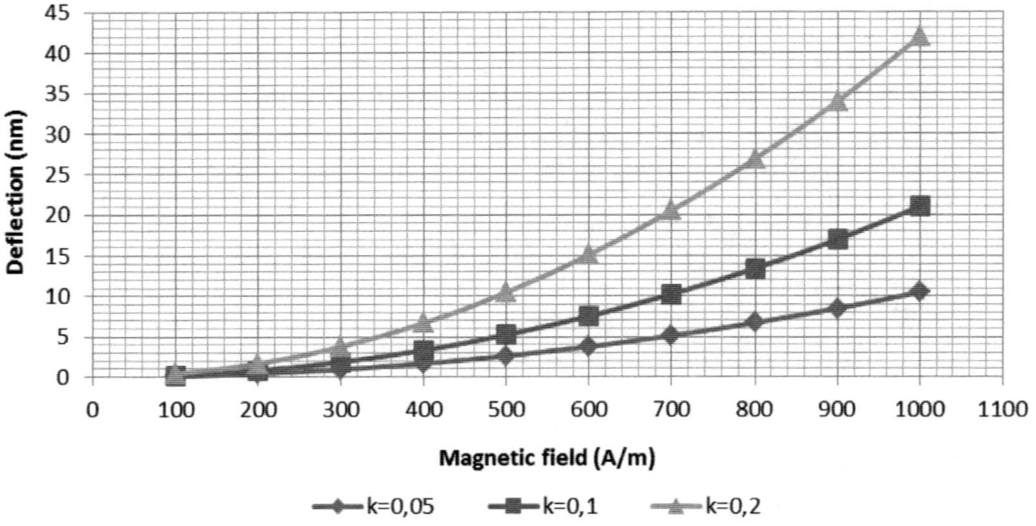

Figure 6. Deflection of the 2.6x2.5 mm cavity cut-out 3x6 mm cantilever construction as a function of magnetic field, for different k factors.

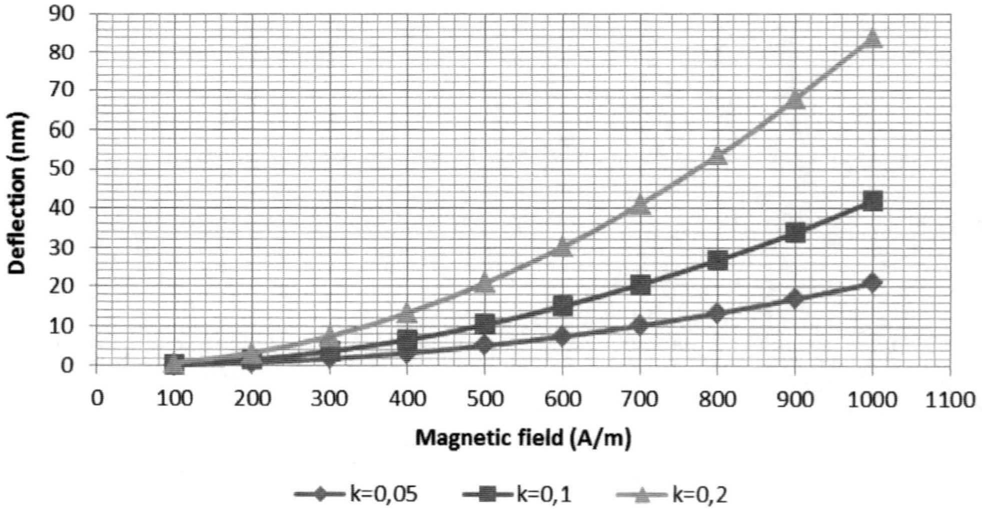

Figure 7. Deflection of 2.8x2.5 mm cavity cut-out 3x6 mm cantilever construction as a function of magnetic field, for different k factors.

Figure 8 shows comparison of results for the two cut-out constructions for the same k factor and varying magnetic field strengths. The increase of width of the cut-out surface effects in much higher deflections (the loss of stiffness). Figure 9 shows combined results achieved for the best construction of the cantilever and the planar coil.

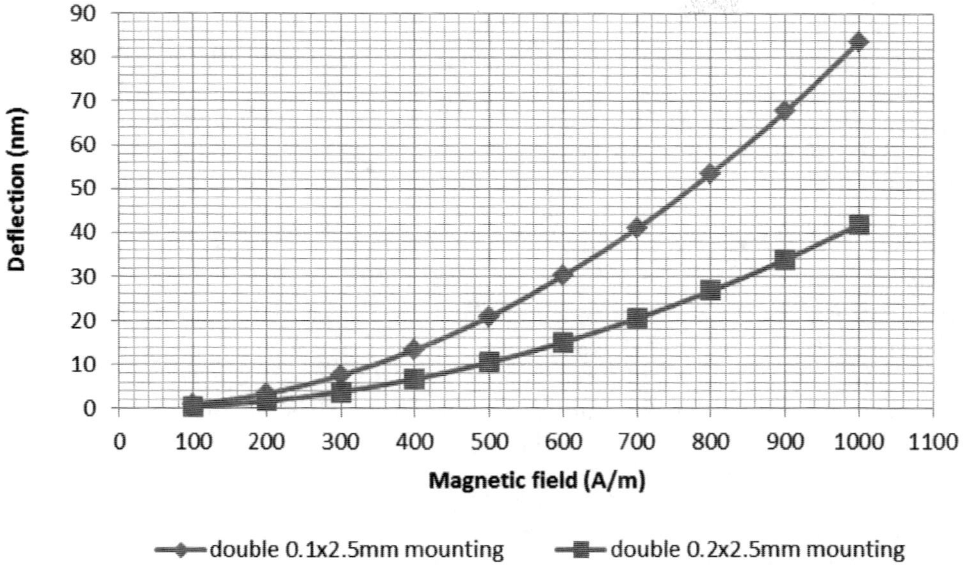

Figure 8. Comparison of deflection of different 3x6 mm cavity cut-out constructions as a function of magnetic field, for k=0.2.

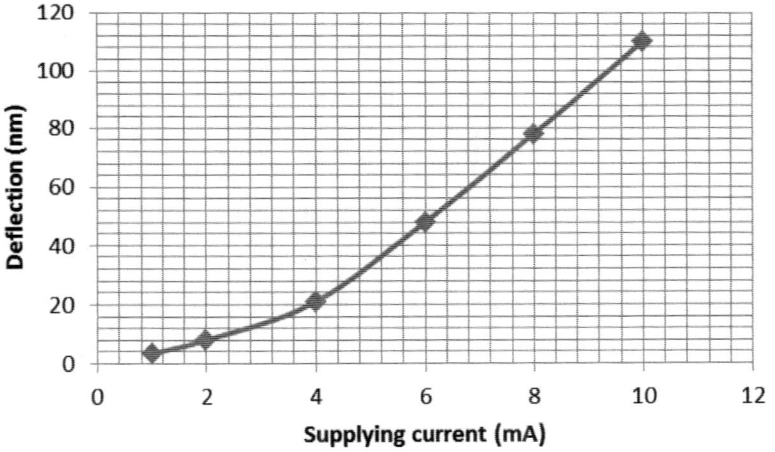

Figure 9. Deflection of the 3x6 mm, 2.8x2.5 mm cavity cut-out construction for k=0.2, as a function of current supplying the planar coil.

4. Conclusion

An influence of the actuators parameters on characteristics was analysed. Comsol Multiphysics 4.3b was used. The study shows the analysis of ferromagnetic layer parameters and beam's dimensions on the optical switch characteristics. Different constructions of the beams were simulated for a range of values of magnetic field strength from 100 to 1000 A/m. The full beams occur to be too stiff to reach reasonable deflections (no more than 1 nm was observed). The loss of stiffness of whole construction, caused by cutting-out a cavity in cantilever's surface effects in much higher displacements, reaching 85 nm for the best construction, even though the volume of the ferromagnetic layer is much smaller (nearly 50% smaller in comparison to the full-cantilever construction). Reducing width of the attaching beams by 50% causes a 50% growth of the whole construction's deflection. The study was carried out for magnetic field strengths occurring in practice [10]. The beams were placed in a very small distance to the planar coil to minimize the losses in the air-gap. However, the planar coil used in the simulation, in real case was able to generate only the lower values of magnetic field strengths, due to need of high supplying currents (the currents reaching 10 mA were used during the simulations). The study proves that deflections of the beams in real-life cases are big enough to be measured by an optical transducer.

5. References

[1] Suhadolnik A , Babnik A , Mozina J 1995 Optical fiber reflaction refractometer *Sensors and Actuators B* **29** 425-432
[2] Golebiowski J and Prohun T *Proceedings of INCINCO Second International Conference on Informatics in Control, Automation and Robotics* (Spain: Barcelona) p 301-306
[3] Binu S, Mahadevan Pillai V P , Chandrasekaran N 2007 Fibre optic target reflectivity sensor *Opt Quant Electron* **39** 747-752
[4] Judy J W and Muller R S 1996 Magnetic Microactuation of Torsional Polysilicon Structures *Sensors and Actuators A* **A53 no. 1-3** 392-397
[5] Judy J W and Muller R S 1997 Magnetically Actuated, Addressable Microstructures *Journal of Microelectromechanical Systems* **6 no. 3** 249-256
[6] Abbot J J, Ergeneman O, Kummer M P 2007 Modeling magnetic torgue and force for controlled manipulation of soft-magnetic bodies *IEEE Transactions on Robotics* **23 no.6** 1247-1252
[7] Golebiowski J, Prohun T, Rybak M 2004 Modelling of the Silicon Membrane Vibrations Generated by Means of Electromagnetic Forces *WSEAS Transaction on Systems* **Issue 7 3** 2538-2540
[8] Golebiowski J 2012 Microelectromechanical transducer with optical readout for magnetic flux density measurements *Electrical Review* **5** 22-26
[9] Golebiowski J and Milcarz S 2012 Numerical Modeling of MEMS Sensor with Planar Coil for Magnetic Flux Density Measurements *Proc. Comsol Confer.* (Italy: Milan)
[10] Ripka P and Tipek A 2007 *Modern sensors handbook* ISTE Ltd ISBN 978-1-905209-66-8

Journal of Physics: Conference Series **494** (2014) 012018 doi:10.1088/1742-6596/494/1/012018

Simulation of Thermal Processes in Superconducting Pancake Coils Cooled by GM Cryocooler

M Lebioda[1] , J Rymaszewski[1] and E Korzeniewska[1]
[1]Institute of Electrical Engineering Systems, Lodz University of Technology, Stefanowskiego 18/22, 90-924 Lodz, Poland

E-mail: marcleb@matel.p.lodz.pl

Abstract. This article presents the thermal model of a small scale superconducting magnetic energy storage system with the closed cycle helium cryocooler. The authors propose the use of contact-cooled coils with maintaining the possibility of the system reconfiguring. The model assumes the use of the second generation superconducting tapes to make the windings in the form of flat discs (pancakes). The paper presents results for a field model of the single pancake coil and the winding system consisting of several coils.

1. Introduction

Superconducting magnetic energy storage systems (SMES) are one of the most interesting and attractive applications of superconducting materials. Energy is stored in the form of magnetic energy generated by the dc current flowing through the superconducting coil. The first concepts from early 1970s assumed the construction of large plants that would store hundreds of megawatt hours of energy [1]. Currently designed SMES are used rather to minimize the effects of voltage fluctuations and are characterized by low dimensions and a small quantity of stored energy [1-4]. The main advantage of SMES systems, compared to the other storage systems, is fast, reliable and flexible power compensation in power systems. Therefore, SMES help to improve the quality and stability of power systems. A typical SMES consists of the following components: the superconducting coil (electromagnet), the power conditioning system (energy converters), the control unit and the cryogenic system [2]. The contact cooling method, used in designed system, imposes numerous restrictions in the design of the coils, current leads and cooling system, but it is the most economical and safe method for temperatures lower than 40 K.

In this article, the authors focused on the modelling of the cooling process of the high-current coils made of the second-generation (2G) ReBCO superconducting tapes. Currently, the winding of the pancake (flat) coils is the most popular method of winding coils from 2G tapes. Depending on the application, superconducting windings are configured in the form of a stack (high magnetic field electromagnets) or the form of a toroid (SMES). ReBCO compounds exhibit a lower anisotropy of the crystal structure and of the electronic properties in comparison to BSCCO compounds. It implies a stronger coupling of vortices to form flux lines, a lower tendency to flux creep effects and a much larger irreversibility field [5]. Preparation methods of ReBCO tapes (depositing, coating) allow to obtain the high-quality, oriented superconducting films with a limited number of the weak links amongst grains. Therefore, 2G tapes offer unprecedented electromagnetic (very high critical current and upper magnetic field) and also mechanical (high tensile strength and mechanical stability) parameters. The great advantage of the 2G tapes is also high thermal stability resulting from the very small thickness of the superconducting ReBCO layer (about 1 μm) in relation to a substantially thicker layer of metallic substrate (tens of μm) [6]. The temperature margin for coils made of high temperature superconducting materials is greater by an order of magnitude in comparison to the low temperature superconductors and the heat capacity is greater by 2-3 orders [7]. Favorable electromagnetic, mechanical and thermal parameters of new generation tapes allow building devices with much smaller dimensions (e.g. integrated with a closed cooling system using helium cryostat). These solutions may be significantly less complicated in comparison to cooling systems with cryogenic liquids. The winding construction should provide a high thermal stability in the static and dynamic states at large currents and a sufficiently mechanical stability, which provides a small degradation of the winding insulation. The authors propose the use of contact-cooled coils with maintaining the possibility of reconfiguring the system.

The initial cooling process plays a key role in the operation of magnet systems cooled by crycoolers [8]. The functionality of the system is significantly determined by the initial cooling time. The paper presents results (dynamics of cooling process) for a field model of the single pancake coil and the winding system consisting of seven coils.

2. Analysis of thermal processes

Analysis of the nature of thermal processes in the vacuum-insulated cryostat was first and crucial stage of the work. Contact cooling requires a special construction of the cryostat, in which the cooled element is placed in a vacuum, which is the main insulation [9]. The heat exchange between the cooled parts and the head of the cryocooler is carried out by heat conduction. A good connection between the heat exchanger and the cooled part and the effectiveness of thermal insulation from the environment are key factors in designing of the cooling system. The factors that may influence the cooling efficiency in this method are:

- maximum heat load of the crycooler, which results directly from the type and construction of the device. The heat load is strongly temperature-dependent and decreases with decreasing temperature;
- heat capacity of the cooled part, which is directly related to the mass of the part and its specific heat capacity, which depends on the amount and type of used material. Copper is the most common used material because of the high thermal conductivity and relatively low price;
- the quality of the connection between the cooled part and the cooling head. Thermal resistance of the connection should be minimized and constant in the wide temperature range (especially at low temperatures). The contact surface should be as large as possible, to reduce the heat flux density. The soft indium foil is used to create "seals" in such cases [10];
- thermal insulation. Vacuum insulation (reducing heat conduction in the gas) and the radiation shields are the most commonly used methods of reducing heat exchange with the environment.

Optimal construction of the cooling system should eliminate uncontrolled heat transfer inside the cryostat, i.e. radiation and convection in the gas. The existence of a small amount of gas in the vicinity of cooling head and cooled parts leads to the occurrence of convection, which can be considered as a combination of the thermal conductivity of the gas and the dynamics of movement of its particles. The reduction of gas pressure in the cryostat greatly reduced heat transfer by convection of gas. This is due to the increase in the mean free path of gas molecules resulting from the reduction of pressure and temperature.

Radiation is the dominant mechanism generating losses in cryogenic systems. Thermal radiation is the energy emitted in the form of electromagnetic waves by a body having a specific thermal energy at a given temperature. The energy transport by radiation does not require a medium, however it plays a dominant role in vacuum cryostats. The use of metal shields of highly reflective material and decreasing the emissivity coefficient are the most effective methods of reducing radiation [11]. The surface quality of the used materials is particularly important. Materials coated with oxides and matt surfaces are particularly disadvantageous.

3. Experiment

The authors intend to design and make the operating, micro superconducting energy storage system. This paper describes the first stage of the project, i.e. developing the concept of the implementation of windings, current leads and the cooling. The system uses the closed cycle helium cryocooler ARS DE-210AF with compressor M600. The applied cryocooler is a two-stage Gifford-McMahon refrigerator, characterized by high refrigeration power (25W at 77 K – on the first stage and 4W at 9K –on the second stage). The workspace in the radiation shield has the diameter of 170 mm and the height of 100 mm.

In the first part of the experiment, the authors have determined characteristics of the cooling process with variety geometries and heat capacities of the heat load. The objective of this study was to analyze the dynamics of cooling. The combinations of two heat exchangers: a copper block (80x80x10 mm) and a copper disc (140x10 mm) were examined. The exchangers were mounted directly on the cold finger of the cryostat. An indium foil was used to seal the connection. The study involved three load configurations: block, plate, block placed on the plate (figure 1).

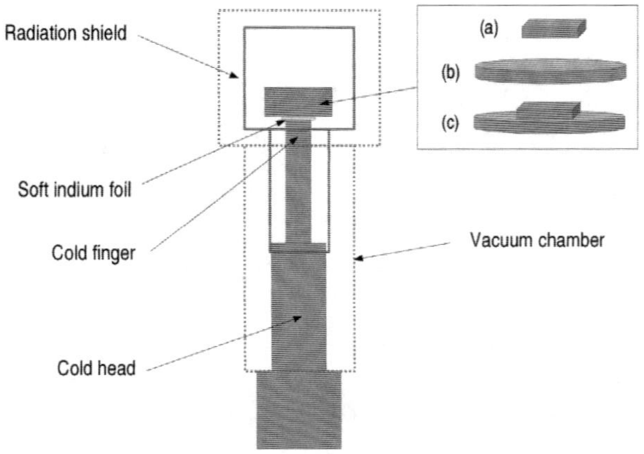

Figure 1. Heat load configurations: (a) copper block, (b) copper plate, (c) block placed on the plate.

The measured time of cooling the exchanger depended mainly on the heat capacity of the entire system and the initial temperature. The cooling time increased together with the increasing of the heat load, but the nature of the process remained unchanged. This indicated a good tightness of the vacuum system and a proper selection of the radiation shields. The process was linear up to approx. 40 K, then it was characterized by a much higher dynamics (figure 2).

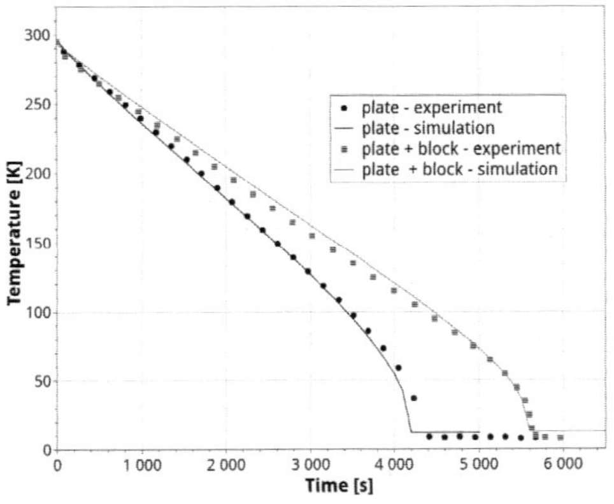

Figure 2. Dynamics of cooling process on cold finger obtained for actual system and determined from model (various configurations of heat exchanger).

This change can be identified with the increase of thermal conductivity of copper, typical for this range of temperatures, and the change of the kinetic energy of gas molecules due to lower temperatures, which transfers into a deceleration of the gas molecules, and means the additional reduction in the gas pressure. The study was carried out in a medium vacuum at a pressure of 60 mTorr and at the initial temperature of 295 K.

The experimental demonstration of the dominant role of radiation as a source of heat loss in vacuum cryostats was the next stage of the research. The plate+block configurations with and without the radiation shield were examined (figure 1c). Only the upper part of the shield covering the exchanger was removed. Temperature of the cold finger and temperature on the surface of the heat exchanger were measured. Removing the radiation shell impeded the cooling process particularly in the low temperature range (figure 3). For the system without the radiation shield the temperature of the exchanger could not be reduced below approx. 85 K. The increase of temperature of the heat exchanger extended the time of cooling the cold finger. This confirms the domination of the radiation process at low temperatures in vacuum cryostats.

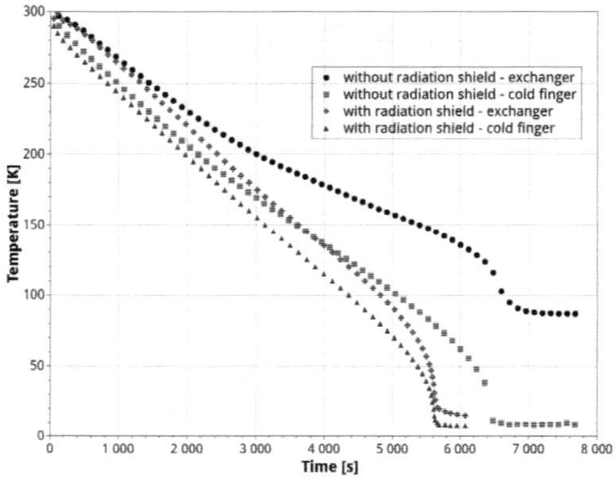

Figure 3. Time dependence of temperature at cold finger and surface of heat exchanger with and without radiation shield.

In the second part of the study the authors have made the prototype of the device with a single superconducting coil (figure 4). The cold finger of cryostat was attached to a heat exchanger in the form of a circular disc (diameter 140 mm, height 10 mm), made of electrolytic copper. The flat superconducting coil (45 turns, outer diameter 105 mm, inner diameter 75 mm) was attached to the heat exchanger.

Figure 4. Prototype device - superconducting coil attached to heat exchanger.

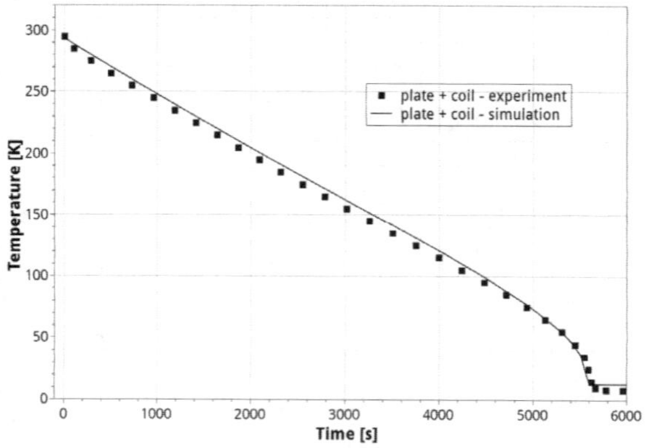

Figure 5. Dynamics of cooling of single coil attached to copper plate obtained from experiment and from simulation (temperature is measured at cold finger).

High heat capacity of the heat exchanger ensured high thermal stability of the operation of the winding. The current leads were made of flat copper bars. The contact surface between leads and the heat exchanger was as large as possible. The solution minimized the amount of heat transported to the coil from the environment through current feedthroughs. Kapton tape of a thickness 50 microns was used for electrical insulation. The superconducting coil with cooled current leads adds small heat load to system. It results from their small heat capacity in relation to the heat exchanger. Thus, the cooling time is only slightly longer (about 5 minutes) in comparison to the cooling time of the heat exchanger (figure 5).

4. Models

Further design work was preceded by the development of thermal and electrical model of the designed device. The model was implemented in COMSOL Multiphysics package [12]. The primary objective of this part of the work was the developing an universal model, accurately describes the thermal processes occurring during the cooling in the cryocooler. Empirical studies carried out for various loads and various geometries were the knowledge base required to develop the model. The model allows to determine the cooling time of components with different geometries and made of different materials and determine the most basic thermal parameters of the analyzed systems. The model also enables an analysis of the dynamics of the process. The possibility of extending the described processes with phenomena related to the electric current flow and the Joule heat generation is an additional feature of the model. The study was divided into two sub-steps. During the first, geometries corresponding to systems tested empirically were examined (figure 1). In the second part the geometry of the superconducting magnet with the pancake coils (part of SMES) was analyzed.

Heat transport is described by the conventional heat equation [12]:

$$gC_p \frac{\partial T}{\partial t} - gC_p \mathbf{u} \cdot \nabla T = \nabla \cdot (k \nabla T) + Q , \tag{1}$$

where g is the density of the solid; C_p is specific heat; T is the temperature, k is thermal conductivity, u is the velocity vector of particles and Q is the heat source. The convection can be neglected for solids ($\mathbf{u}$=0), therefore equation (1) can be converted to the following form [12-14]:

$$gC_p \frac{\partial T}{\partial t} - \nabla \cdot (k \nabla T) = Q . \tag{2}$$

The Q term represents Joule heating and can be described as follows:

$$Q = \mathbf{E} \cdot \mathbf{J} , \tag{3}$$

where $\mathbf{E}$ is the electric field strength (V – electric potential):

$$\mathbf{E} = -\nabla V , \tag{4}$$

$\mathbf{J}$ is the total current density (ε – the electric permittivity, $\mathbf{J_e}$ – externally generated current density) [11]:

$$\mathbf{J} = \left(\frac{1}{\rho} + \varepsilon \frac{\partial}{\partial t} \right) \mathbf{E} + \mathbf{J_e} . \tag{5}$$

All the parameters of the materials used in the model are temperature dependent (figure 6). Taking into account the effect of the temperature on materials parameters is necessary to preserve the accuracy of the model. It results from the strong influence of temperature on the parameters of copper in the range T <100K (increase of the conductivity and decrease of the specific heat – figure 6). Moreover, copper is the predominant material in 2G superconducting tapes (over 50% of volume) and is a mechanical and thermal stabilizer for superconducting core. Therefore, the model can ignore the impact of temperature on the parameters of the superconducting core and buffer layers, however, it is necessary to take into account the temperature dependence of physical parameters of copper and Kapton insulation. Adequately to the experimental studies it was assumed that heat exchangers are made of copper. The model takes into account two mechanisms of heat exchange with the environment: convection in the gas at reduced pressure and radiation. The boundary conditions for the thermal model (the heat flux on boundaries) are defined as follows [12]:

$$-\mathbf{n}(-k\nabla T) = q_0 + h(T_{inf} - T) + e\sigma(T_{amb}^4 - T^4), \tag{6}$$

where **n** is the unit vector normal to a boundary surface, h is the heat transfer coefficient, q_0 is the inward heat flux, T_{inf} is the ambient bulk temperature, T_{amb} is the temperature of the surrounding radiation environment, e is the emissivity coefficient, σ is the Stefan–Boltzmann constant. Convection cooling is modeled by Newton's law of cooling $h(T_{inf} - T)$. The value of the heat transfer coefficient depends on many factors, i.e. the surface temperature, properties of the cooling fluid (e.g. pressure), the fluid's flow rate and also the geometrical configuration. The general form of the coefficient h can be written as follows [12]:

$$h = \frac{L}{k}\mathrm{Nu}(\mathrm{Ra}, \mathrm{Pr}, \mathrm{Re}), \tag{7}$$

where L is the characteristic length (e.g. length of wall); Nu, Ra, Pr, Re – characteristic dimensionless numbers: Nusselt, Rayleigh, Prandtl, Reynolds. The characteristic numbers depend on the of convection conditions and are described by, often complex, empirical and theoretical correlations. The phenomenon of the natural convection was taken into account using Comsol module, which allows to determine the convection coefficient based on the shape and dimensions of the elements (for vertical and horizontal walls) and the pressure of the surrounding gas. The pressure value was obtained from experimental studies (60 mTorr). However, in described case the heat transfer is mainly determined by the radiation mechanism. The model assumes that the heat emissivity coefficient e of the shield is constant and equals to 0.07 (polished Cu).

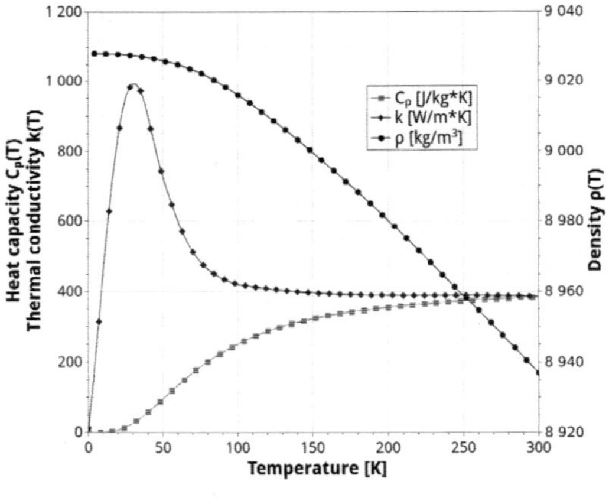

Figure 6. Temperature dependencies of thermal properties of copper: $C_p(T)$ - heat capacity [15], $k(T)$ - thermal conductivity [16,17], $\rho(T)$ - density [16,18].

The connection between the heat exchanger and the cold finger is also included in the boundary conditions. The heat transfer at the connection is described as a heat flux directed from the heat exchanger through the surface corresponding to the contact with the cold finger (q_0 is negative). The heat power transferred from the heat exchanger is described by the thermal dependence of the cryocooler heat load, provided by the manufacturer (figure 7) [19]. This assumption was necessary because the characteristics of the heat load is determined for the load applied to the cold finger and does not consider the heat capacity of the cryocooler elements as a separate factor. Thus, the geometry of the cold finger cannot be included in the model. The result of this assumption is inability to take into account an effect of the quality of the connection between the heat exchanger and the cold finger.

Thermal resistance of the connection produces a difference between the temperature of the finger and the temperature of the heat exchanger (figure 3).

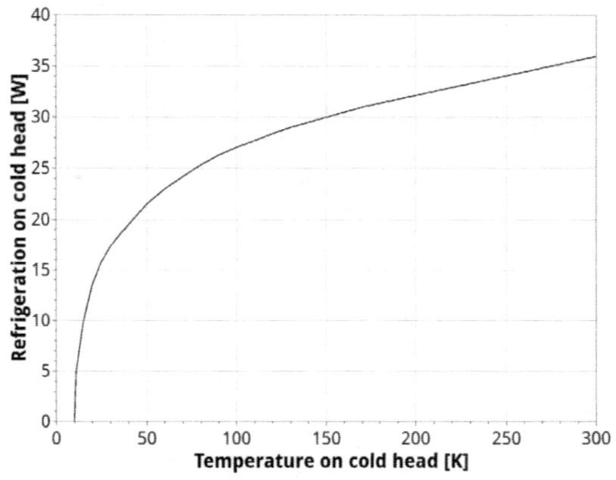

Figure 7. Map of heat load on the second stage of cold head of DE210A cryostat [19].

The electrical boundary conditions assume electric insulation at all boundaries, with the exception of the single surface of current leads. The current flows into the first lead. The electric potential at the second lead equals zero.

The model is solvable despite the relatively complex geometry (figures 1, 8, 9), e.g. for a model shown in figure 9 the number of elements is 184412, the number of boundary elements is 54651. Calculations were performed using the iterative multigrid solver and FGMRES method. The computation time for the model with temperature-dependent parameters (analyzed period - 0…80000 s) was 2382 s, for the model additionally taking into account the current flowing was 4379 s (Intel(R) Xeon(R) X3440 with 12 GB RAM). The main objectives of modelling were the investigation of the cooling process, the comparison with the actual results and the assessment of the reasonableness of further development of the project.

Figures 2 and 5 show the comparison of the dynamics of the cooling process for the actual systems with various configurations of the heat exchanger (including a coil) and the results obtained for the numerical model (temperature 20K is reached after approx. 80 minutes). The simulation results are consistent with the results of the experiment. It confirms the validity of the used assumptions.

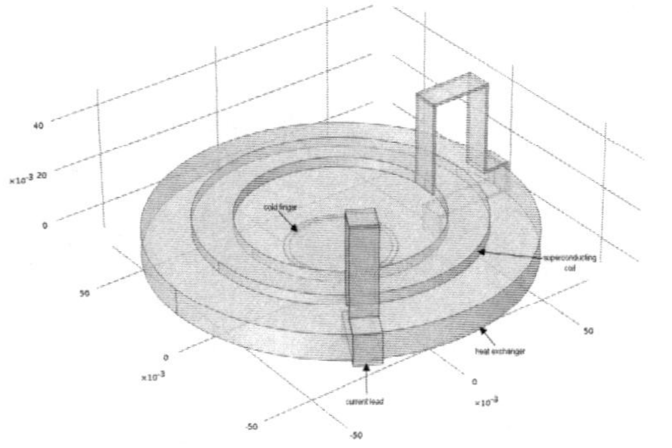

Figure 8. Geometry of simple model with single coil.

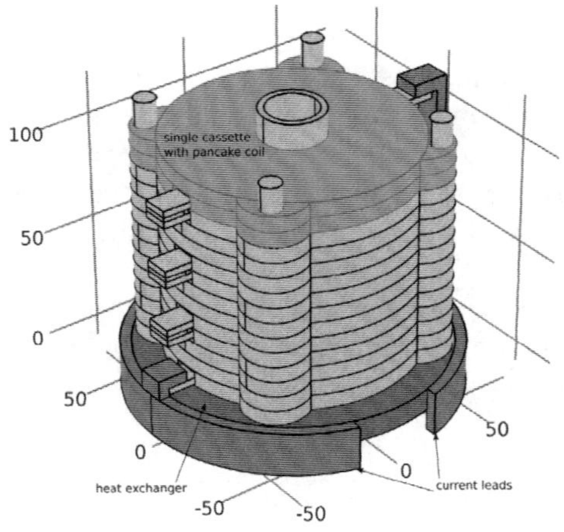

Figure 9. Geometry of SC-magnet model with seven coils (single cassette selected).

Basing on the model with a single coil (figure 8), an extended model with seven coils was developed in the next step (figure 9). The model assumes that the pancake coils (130 turns) are wound from 2G superconducting tapes with the copper stabilizer of a width of 12 mm. In order to improve cooling efficiency and ensure the mechanical stability the coils were placed in the copper cassettes. The windings are configured in a stack and placed on a heat exchanger (copper plate) with the parameters given in the previous model. The designed system is adapted to work with the current of about 500 A and requires a different design of the current leads with increased cross-section. The proposed high current leads are cooled by the contact method with the side surfaces of the heat exchanger (figure 9).

Figure 10 shows the temperature distribution in the cross-section of the device, obtained after a period of approximately 11 h. The presented distribution is uniform throughout the entire volume of the windings (21 K). It confirms the suitable thermal stability of the system.

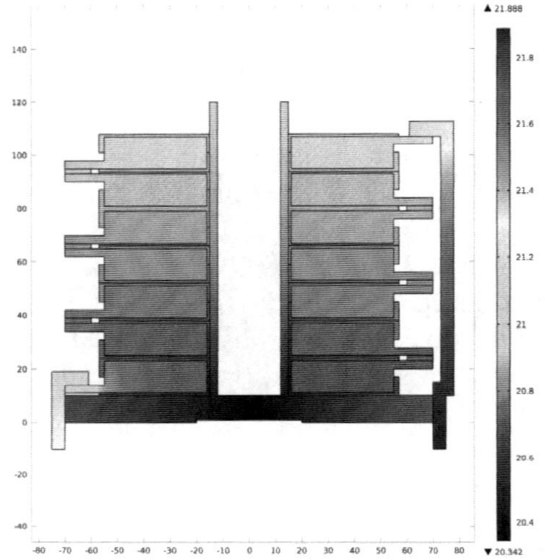

Figure 10. Temperature distribution after approx. 11 hours of cooling ($I = 0$ A).

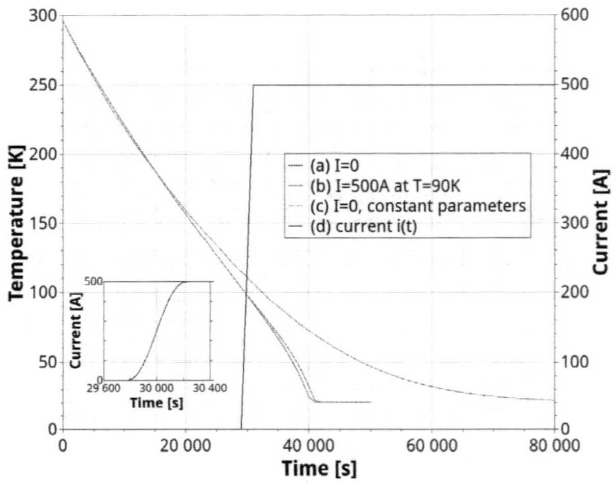

Figure 11. Dynamics of cooling of SC-magnet (temperature is measured at cold finger) and current-time characteristic.

Figure 11 presents time dependency of the temperature at the cold finger for SC-magnet. The model assumes that the current in windings increases with a small slope (di/dt = 1 A/s) (figure 11d), therefore the losses resulting from eddy currents are negligibly small. The coil current causes Joule heating and is the main reason of extending the cooling time (figure 11b). In addition, the change of material properties (figure 6) and the crycooler heat load (figure 7) influence on the process. In order to investigate the effect of variability in material parameters the simulations for constant values were carried out. In this case, the simulated cooling time significantly increased and the nature of the process was changed (figure 11c). This result is inconsistent with the experimental results. The time of current switching is the most important in the cooling of magnets. It confirms the crucial role of the initial cooling in the operation of the magnet [8].

5. Summary

Application of 2G superconducting tapes to build superconducting magnetic energy storage systems can reduce device size while maintaining operating parameters. Large usable critical current density and high thermal stability of these tapes allow for the construction of small-size devices, which promotes the use of cryocoolers and contact cooling - simple in design or operation and economical refrigeration method. The main distinction of cryocoolers is that cooling is provided by a closed loop system, which eliminates the cost of the continuous supply liquid helium. Contact cooling in contrast to cryogenic cooling in liquid ensures full control over the operating temperature of the superconducting device. In the case of superconducting magnetic energy storage, it transfers into the amount of accumulated energy.

A significant reduction in the temperature of the coils, below the critical temperature of the superconductor (93 K for 2G YBCO tapes), increases the current density of the superconducting tape, and consequently increases the capacity of the energy storage. It also significantly increases the thermal stability and the safety margin of the windings, which is especially important in dynamic states of high-current devices. A comparison of temperature dependencies of the relative value of the critical current and the magnetic energy shows that cooling of the windings below 20 K will not substantially increase the amount of the accumulated energy.

The presented model correctly describes the thermal processes occurring in the cryostat with vacuum insulation. The obtained results are consistent with the results of physical experiments carried out on different shapes and different volume of cooled elements. It confirms the versatility of the proposed description and the possibility of using the model to describe the thermal processes occurring during the cooling of components with any geometry including SMES. It should be noted that the parameters used in the model, such as gas pressure, the emissivity of the shield, and others correspond to the physical experimental system. Model has not been validated for other configurations.

Well defined thermal model enables qualitative and quantitative analysis of cooling process and optimization of the device design. Experimental verification of the model with a single coil confirms the validity of the assumptions. The presented model is the basis for constructing more complex models of superconducting systems cooled by cryocoolers. For a model with seven coils resulting time to cool down to the temperatures about 20 K is much longer (over 11 hours). Therefore, the presented model requires further optimization, especially reducing the volume of stabilizer.

References

[1] Buckles W and Hassenzahl W V 2000 *IEEE Power Engineering Review* **20** 16-20
[2] Xue X D Cheng K W E and Sutanto D 2006 *Supercond. Sci. Technol.* **19** 31-9
[3] Ali M H, Wu B and Dougal R A 2010 *IEEE Transactions On Sustainable Energy* **1** 38-47
[4] Luongo C A, Masson P J, Nam T, Mavris D, Kim H D, Brown G V, Waters M and Hall D 2009 *IEEE Transactions On Applied Superconductivity* **19** 1055-68
[5] Krabbes G, Fuchs G, Canders W R, May H, Pałka R 2006 *High Temperature Superconductor Bulk Materials: Fundamentals, Processing, Properties Control, Application Aspects* (Wiley-VCH)
[6] Sumption M D, Majoros M, Susner M, Lyons D, Peng X, Clark C F, Lawless W N and Collings E W 2010 *Supercond. Sci. Technol.* **23** 075004
[7] Ishiyama A, Ueda H, Aoki Y, Shikimachi K, Hirano N and Nagaya S 2011 *IEEE Transactions On Applied Superconductivity* **21** 2398-401
[8] Choi Y S, Kim D L, Shin D W 2011 *Physica* C **471** 1440-4
[9] Barron R F 1999 *Cryogenic Heat Transfer* (Taylor & Francis Group)
[10] Salernoa L J, Kittela P and Spivak A L 1994 *Cryogenics* **34** 649-54
[11] Barron R F 1985 *Cryogenic Systems* (Oxford University Press Inc)
[12] COMSOL Multiphysics, Modeling and Simulation Software http://www.comsol.com/
[13] Rymaszewski J, Lebioda M and Korzeniewska E 2011 *Materials Science and Engineering* B: *Solid-State Materials for Advanced Technology* **176** 334-9
[14] Lebioda M and Rymaszewski J 2013 *Przeglad Elektrotechniczny* 89 280-3
[15] White G K and Collocott S J 1984 *J. Phys. Chem. Ref. Data* **13** 1251-7
[16] Simon N J, Drexler E S and Reed R P 1992 *Properties of copper and copper alloys at cryogenic temperatures* (National Institute of Standards and Technology, U.S. Department of Commerce)
[17] Ho C Y, Powell R W and Liley P E 1972 *J. Phys. Chem. Ref. Data* **1** 279-421
[18] Hahn T A 1970 *J. Appl. Phys.* **41** 5096-101
[19] Advanced Research Systems http://www.arscryo.com/

Journal of Physics: Conference Series **494** (2014) 012019 doi:10.1088/1742-6596/494/1/012019

Modelling of MOCVD Reactor: New 3D Approach

E Raj[1,3], Z Lisik[1], P Niedzielski[1], L Ruta[1], M Turczynski[1], X Wang[2], A Waag[2]
[1] Lodz University of Technology, Department of Semiconductor and Optoelectronic Devices, Wolczanska 211/215, 90-924 Lodz, Poland
[2] Institute of Semiconductor Technology, TU Braunschweig, Hans-Sommer-Strasse 66, 38106 Braunschweig, Germany

[3]E-mail: ewa.raj@p.lodz.pl

Abstract. The paper presents comparison of two different 3D models of vertical, rotating disc MOCVD reactor used for 3D GaN structure growth. The first one is based on the reactor symmetry, while the second, novel one incorporates only single line of showerhead nozzles. It is shown that both of them can be applied interchangeably regarding the phenomena taking place within the processing area. Moreover, the importance of boundary conditions regarding proper modelling of showerhead cooling and the significance of thermal radiation on temperature field within the modelled structure are presented and analysed. The last phenomenon is erroneously neglected in most of the hitherto studies.

1. Introduction

Metal-Organic Chemical Vapour Deposition (MOCVD) is one of the most widely used epitaxial technologies to manufacture electronic devices, especially for optoelectronic applications. Even though, this thin-film deposition technique is used for mass production, it is highly complex process involving reacting flow of organometallic and hydrides precursors accompanied by carrier gasses [1–4]. Its complexity is especially intricate in the case of three-dimensional structure growth, like the 3D GaN column structures destined to solid state lighting. This application is in the scope of authors interest. The pillar design of LEDs allows getting even more than 10 times bigger active area in comparison to the planar solutions [5–7]. Moreover, the vertical alignment of junctions assures lower influence of thermal expansion mismatch between the columns and the substrate, as well as it results in low defect concentration regardless of the material type. These benefits cause that 3D LEDs and their production techniques, with special emphasis on MOCVD, gain more and more attention of various research groups, nowadays [5–13]. The studies are oriented on understanding the selective growth of semiconductor materials [5–8], as well as on experimental and numerical analyses of MOCVD parameters and their influence on the growth processes [9–13].

There are several constructions of MOCVD reactors with both lateral and vertical gas flow [14] but from the point of view of optoelectronic applications more popular are close coupled showerhead reactors [5-7,10]. The outline of such a reactor is presented in figure1. The reactants are introduced through a showerhead located over a susceptor and flow down towards its surface, where the film deposition process occurs. The susceptor is in a form of the rotating, heated disc, while the showerhead is water cooled, so huge temperature differences can be observed inside the chamber. The forced flow of the carrier gasses and the precursors may be influenced by the buoyancy force and the rotational motion of the susceptor. As the result, the three-dimensional velocity field inside the reactor

Journal of Physics: Conference Series **494** (2014) 012019 doi:10.1088/1742-6596/494/1/012019

is observed, what is crucial from the point of view of modelling. Moreover, the chemical reactions are mostly temperature activated, so the knowledge of the temperature field inside the chamber with special emphasis on the processing area is a must to properly analyse the growth mechanism.

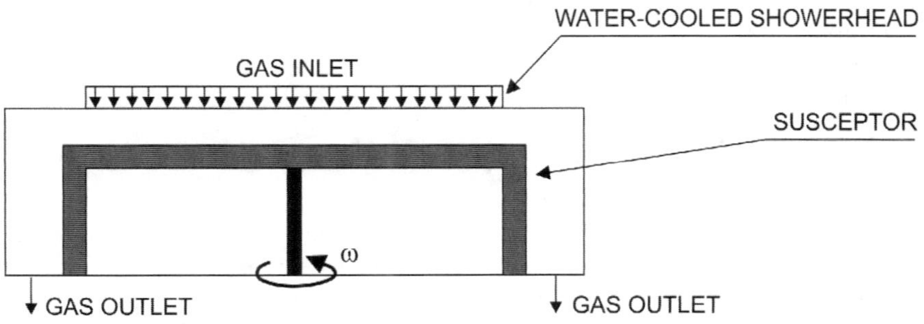

Figure 1. Outline of MOCVD vertical, rotating disc reactor.

Although, MOCVD reactors have been under numerical investigations for several decades already [1–13], the constructions of particular reactors differ significantly and the differences of elaborated models are significant and dependent on type, shape and dimensions of analysed chambers. Fotiadis et al. [9] studied vertical rectors under wide range of operating parameters such as susceptor and chamber walls temperatures, pressure, inlet flow rate and susceptor rotational speed. The test reactor had 140 mm in diameter susceptor and was characterised by the high distance between the gas inlet and the processing area changing in the range from 40 to 240 mm. The conducted analyses showed that the reactor geometry is of crucial importance to reduce buoyancy induced vortexes and to achieve high uniformity of thin-film layer deposition. Furthermore, Zhong et al. [10] conducted 2D simulations of the similar reactor but with much smaller distance between the susceptor and the showerhead equalled to 10 mm. They concentrated on transient flows and the possibility of temporary oscillations that might increase the growth rates of semiconductor compounds and might result in structure inhomogeneity. On the contrary, Mitrovic et al. [11, 12] focused on high-speed rotating MOCVD reactors characterised by susceptors with diameters of 300 and 450 mm and the distance between the susceptor and the showerhead of 75 and 125 mm, respectively. Their "pressure-rotational rate" diagrams allows predicting the influence of the carrier gases and reactants flow on the growth rate under wide range of varying conditions in big chambers, when the rotational speed remarkably exceeds 100 rpm.

In general, one can conclude the 3D models are necessary to accurately describe the temperature and the velocity fields inside the disc rotating MOCVD reactors. On the other hand, the questions regarding importance of different phenomena are still raising. For example, in the most of papers [8, 10–13], the radiation heat transfer mechanism is neglected, even though some authors in their early studies [9] state clearly that such an attitude is not allowed in MOCVD reactors. Furthermore, the showerheads are usually represented as a uniform velocity boundary condition [1,2,9–13] and the cooling of either showerhead or chamber walls is implemented in the form of constant temperature boundary condition [10–13]. These assumptions can, however, influence the temperature field in the reactor chamber and may affect the precursors decomposition and the modelled growth process [1–4].

The paper presents the new, reduced 3D model of empty MOCVD reactor used further by the authors to investigate the processes of 3D GaN structure growth. The worked out model allows reducing the usage of computational resources, simultaneously assuring high accuracy of the conducted simulations. Its advantages have been proven in the numerical investigations dealing with the influence of different boundary conditions and assumptions on the temperature and velocity distribution within the considered MOCVD chamber. The carried out investigations are comprehensive taking into account previous studies [8–13].

2. Reduced model of MOCVD reactor

The MOCVD reactor under consideration is a close coupled showerhead reactor of Thomas Swan Scientific Equipment Ltd. that cross-section is schematically presented in figure 1. A graphite susceptor with the diameter of 140 mm is located in 175 mm in diameter, cylindrical chamber, that is closed from the upper side by a water-cooled showerhead. The shower itself contains almost 2000 tubes covering in total the area of almost 135 cm^2. The diameter of each nozzle is 0.6 mm. The distance between the susceptor and the showerhead equals to 11 mm. The outlet is located around the susceptor as it is marked in figure 1.

2.1. Models: assumptions and boundary conditions

Due to the distribution of the tubes at the area of the showerhead, the symmetry has been found and a model covering a quarter of the reactor chamber has been elaborated. The outline of the model, named Q-type, is shown in figure 2a. The complexity of the geometry incorporating matrix of tubes as well as planned introduction of processed samples with 3D GaN structures into the simulations, have forced us to work on much simpler model. The requirement to have a 3D analysis and the design of intricate matrix of showerhead tubes lead us to the second model, named SR-type solution. It is presented in figure 2b, and it covers only a single row of nozzles sliced from Q-type model. In both the cases, the susceptor is marked in red and represents a uniform graphite layer coated with 100 μm silicon carbide.

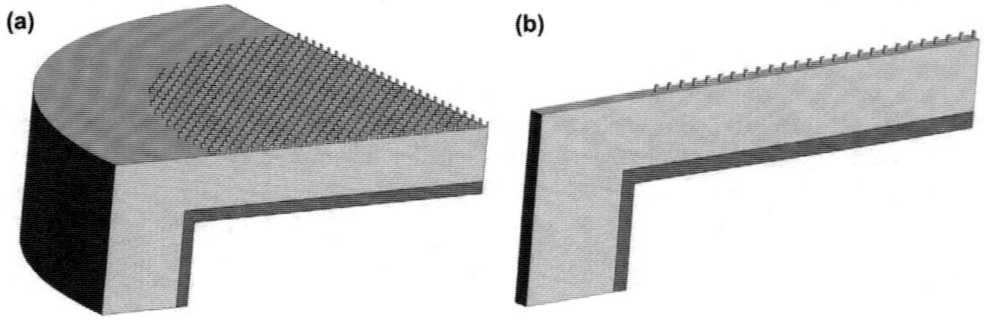

Figure 2. The geometry outline for the model covering a) the quarter of the reactor chamber (Q-type) b) the single row of showerhead tubes (SR-type).

In order to examine the models, the numerical analyses have been conducted with the aid of ANSYS CFX software with the dense mesh covering 5 to 8 inflation layers at every solid surface. Moreover, within the processing area, the first layer thickness has been limited to several micrometers to assure fine grid and an accurate modelling of the main area of interest. It was verified that further refining of the discretisation mesh does not influence the velocity and the temperature fields within the modelled structures.

The simulations with both the models have been conducted for the following assumptions and boundary conditions:

- carrier gas is 100% H_2 (the properties of the fluids are described with the aid of ideal gas law);
- reference pressure in the chamber is 25 kPa and the outlet relative pressure equals to 0 Pa;
- uniform temperature of 1000°C at the bottom area of the susceptor is assumed;
- uniform temperature of 50°C at the tube inlets and at the surface of showerhead (water cooling) is applied;
- radiation thermal model with 0.2 emissivity of various walls bounding the gas flow is incorporated (no heat absorption or scattering within fluid domain has been analysed);
- uniform velocities at the inlets of showerhead tubes ranging from 3000 ÷ 12000 sccm are assumed;

- both stationary solution and the rotation of the susceptor are investigated (rotational speed ω equals to 100 rpm).

The values of linear velocities at the nozzle inlets are calculate from the volumetric flow rates according to the formulas:

$$V_Q = \frac{Q_X}{60 \cdot N \cdot A_{TUBE}} \tag{1}$$

$$V_{SR} = V_Q \frac{N}{2\pi \cdot N_{SR}^2} = \frac{Q_X}{60 \cdot (2\pi \cdot N_{SR}^2) \cdot A_{TUBE}} \tag{2}$$

where: V_Q and V_{SR} are linear inlet velocities for Q and SR-solution respectively [m/s]; N – number of tubes in the showerhead [-]; N_{SR} – number of tubes in SR-model showerhead [-]; A_{TUBE} – cross-sectional area of the single tube in the showerhead [mm^2]; Q_X is a nonstandard volumetric flow at temperature T_X and pressure p_X calculated according to the formula:

$$Q_X = Q_{STD} \frac{p_{STD}}{p_X} \frac{T_X}{T_{STD}} \tag{3}$$

where: Q_{STD} is a standard volumetric flow [sccm] at the standard conditions p_{STD} and T_{STD} of 103 kPa and 273 K;

2.2. Results and discussion

The most important information that must be obtained from the numerical results is temperature distribution within the processing area and velocity profiles in the modelled chamber. Hence, the results have been analysed along several lines located along the susceptor radius at the susceptor surface and over it. The lines and their descriptions are gathered in table 1, and the exemplary ones are presented graphically in figure 3.

Table 1. Description of the lines along which the results are gathered.

NAME	LOCATION
ALONG	along the radius of the susceptor on its surface
500 um	500 μm over the line ALONG
1 mm	1 mm over the line ALONG
3 mm	3 mm over the line ALONG
5 mm	5 mm over the line ALONG

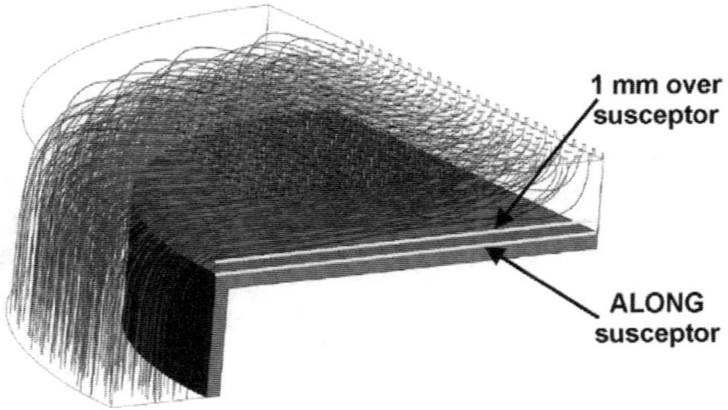

Figure 3. Streamlines in Q-model with two marked lines along which the results are gathered in SR-type and Q-type MOCVD models.

Verification of the reduced 3D model (SR-solution) has covered the comparison of temperature and velocity fields obtained with the aid of both models. Exemplary velocity and temperature distributions versus the distance from the susceptor origin along the four lines defined in table 1 are shown in figure 4 and figure 5, respectively. The results have been obtained for the highest analysed volumetric flow rate of 12000 sccm, and the stationary susceptor.

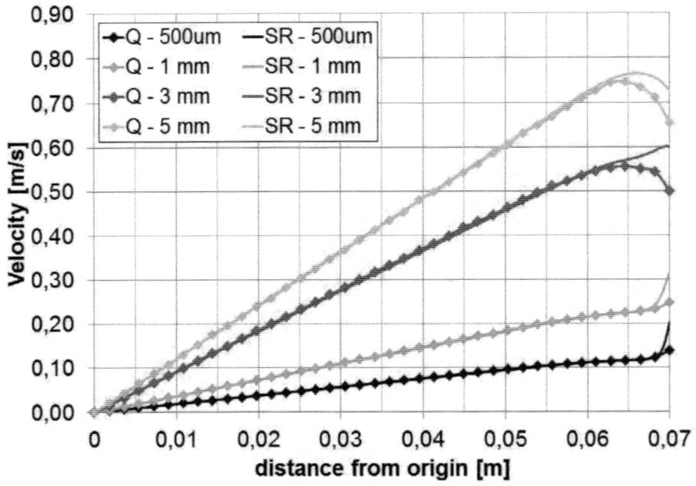

Figure 4. Velocity distribution along four lines in SR-type and Q-type MOCVD models.

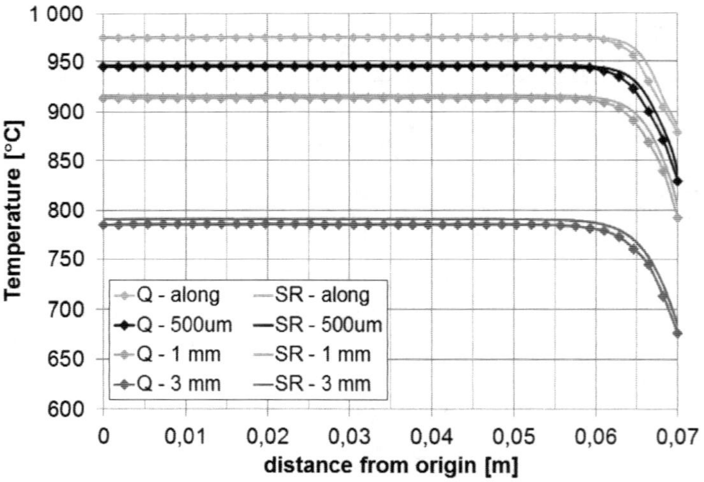

Figure 5. Temperature distribution along two four lines in SR-type and Q-type MOCVD models.

Conducted simulations shown a good conformity among Q-type and the reduced models. The observed differences in the temperature and velocity fields are located outside the processing area and they should have no influence on the growth conditions. Both models allow analysing heat and mass transfer phenomena inside MOCVD reactor with respect to the different operating conditions (type of carrier gases, flow rates pressure etc.).

Furthermore, the impact of the susceptor rotation has been analysed. The simulations have been repeated for both types of the models with implemented domain rotation for the whole analysed range of inlet velocities. The exemplary results along four lines obtained for the highest volumetric flow rate are presented in figure 6, figure 7 and figure 8, respectively. Figure 6 shows lateral (V_X – along the susceptor radius, V_Y – along the susceptor surface perpendicular to its radius) and vertical (V_Z – perpendicular to the susceptor surface) velocity components, while the resultant vector magnitude is illustrated in figure 7. The velocities are given for the coordinate system that refers to the susceptor as a stationary domain. Figure 8 presents temperature distributions. For clarity, the figures do not show the results obtained for the Q-type model.

The susceptor rotation introduces the lateral velocity component perpendicular to the susceptor radius, hence it influences the velocity field in the chamber. It should be emphasised that the V_X and V_Z components distributions obtained for the two analysed cases with and without rotation are comparable. The characteristics almost cover with each other. Furthermore, the 3D flow almost does not affect the temperature field within the tested range. The obtained temperature uniformity in the processing zone is high and the observed differences are located in the area outside the samples' location (beyond 0,06 m) and they are negligible (below 4°C). Nevertheless, the implementation of rotation is a must to predict properly the growth kinetics inside the reactor. The results gained with the aid of Q-type model with implemented rotation are in good conformity with the characteristics shown in figure 6, figure 7 and figure 8; hence, the two presented models can be applied interchangeably.

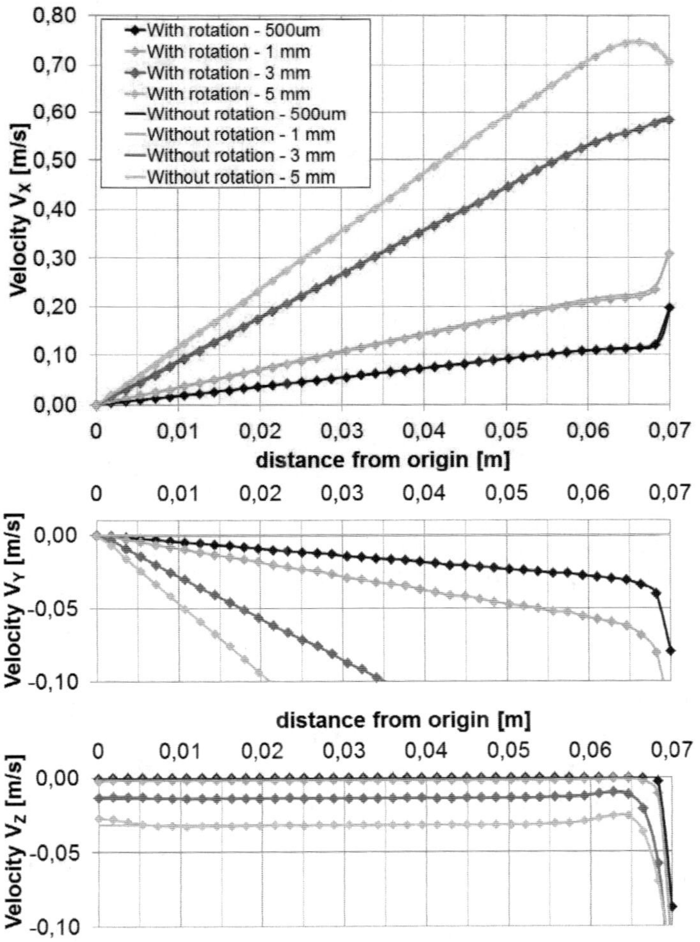

Figure 6. Lateral and vertical velocity components distribution along four lines in SR-type model with and without susceptor rotation.

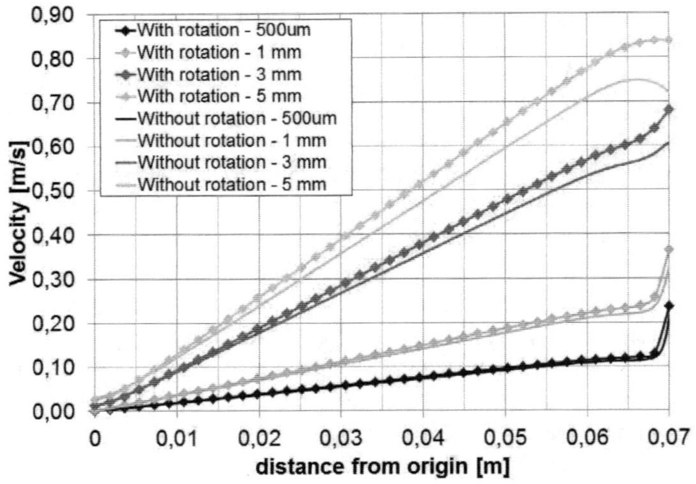

Figure 7. Velocity distribution along four lines in SR-type model with and without incorporated susceptor rotation.

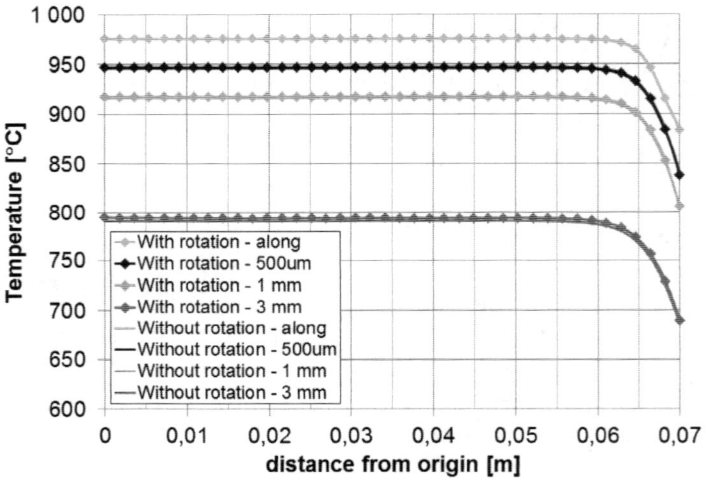

Figure 8. Temperature distribution along four lines in SR-type model with and without incorporated susceptor rotation.

3. Different boundary conditions

Next, with the aid of the reduced SR-type model, the verification of various boundary conditions representing the showerhead cooling has been conducted. In comparison to the previously used assumptions, the constant temperature at the surface of the showerhead has been replaced among simulations by three versions corresponding to different heat exchange between the chamber and the surrounding:

- adiabatic boundary condition named "Adiabatic" – heat is transported only by the cold gases flowing through the reactor chamber,
- constant temperature boundary condition of 50°C named "With T=const" – it represents ideal cooling of the showerhead,
- convective boundary condition of 1000 W/m^{2}K with reference temperature of 50°C named "With Conv." – it characterises the real water cooling systems.

Exemplary temperature distributions for adiabatic, convective and constant temperature boundary conditions obtained for the highest examined velocity are shown in figure 9. To improve readability, the number of characteristics is reduced to the results gathered along the susceptor and along the line located 500 μm over the previous one.

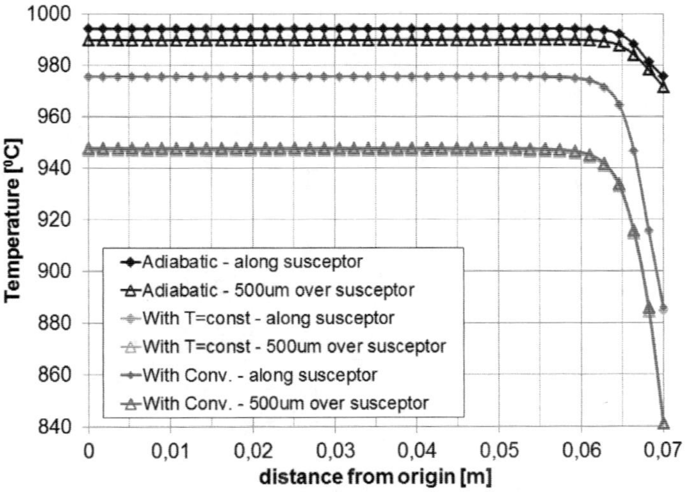

Figure 9. Temperature distribution along two lines in SR-type model for different boundary conditions at the susceptor surface.

Firstly, adiabatic boundary condition results in lower temperature gradients within the modelled chamber. Incorporation of reactor cooling with the aid of only impinging cold gases gives poorer heat transfer rate and it does not describe properly the water-cooled showerhead of the investigated MOCVD reactor. Secondly, both, the constant temperature and the convective boundary conditions allow to receive comparable temperature distribution within the processing area. The difference between the temperature gradients at 1 mm distance from the susceptor surface, obtained for the two cases, does not exceed 1.5°C. Moreover, the average temperature at the showerhead surface for the convective boundary condition is only about 45°C higher in comparison to the constant temperature value of 50°C assigned at this boundary. This difference should not affect the growth process. In consequence, the constant temperature assumptions is proper and allows for adequate modelling of MOCVD reactors with cooled showerheads.

4. Thermal radiation

The last investigated and hereby presented problem is an influence of the radiation transport mechanism on the heat exchange process within the MOCVD reactor. The series of numerical simulations has been conducted mainly for the SR-type model for the all the assumptions and the boundary conditions defined in section 2 except thermal radiation model. Hence, regarding the thermal radiation the three cases have been analysed and they are presented below:

- thermal radiation model is switched off,
- thermal radiation with 0.2 emissivity of various walls bounding the gas flow is incorporated,
- thermal radiation with 0.2 emissivity of walls of MOCVD chamber and 0.8 emissivity of graphite susceptor coated with SiC layer is implemented.

Exemplary results obtained for the highest and the lowest analysed velocity for the model without and with thermal radiation are shown in figure 10. Furthermore, figure 11 presents the solution incorporating thermal model with different emissivity of the susceptor surface. Lower value of 0.2 corresponds to the surface emissivity of all the remaining walls bounding the chamber that can be made of either quartz or stainless steel. Higher emissivity of 0.8 characterises the properties of the graphite susceptor with SiC coating. Similarly, as previously, temperature distributions are presented along the radius line of the susceptor and along the line located 500 μm over the previous one.

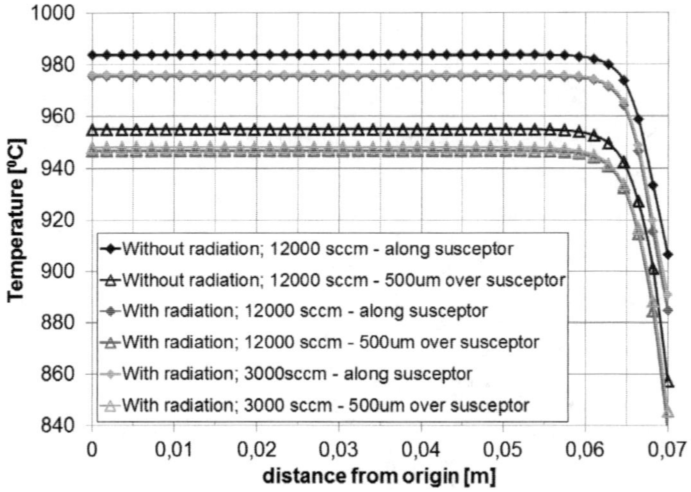

Figure 10. Temperature distribution along two lines in SR-type model: constant temperature boundary conditions with thermal radiation model switched on (0.2 emissivity) and off for two inlet velocities.

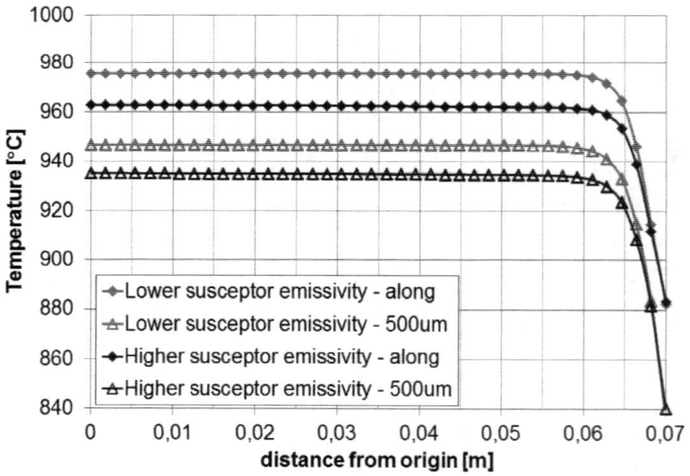

Figure 11. Temperature distribution along two lines in SR-type model: constant temperature boundary conditions with thermal radiation model and different emissivity values of the susceptor.

The curves presented in figure 10 show that both, inlet velocity and thermal radiation influence the obtained temperature distribution within the processing area, even though the impact on the temperature field of the first parameter is insignificant within the tested flow rates. Incorporation of radiation heat transfer mechanism induces higher heat exchange rates, that are further increased by higher emissivity values of the susceptor surface. For the constant temperature boundary condition at the bottom surface of the susceptor equalled to 1000°C, the temperature at the susceptor, where the growth process takes place, is more than 8°C lower for 0.2 emissivity and about 21°C lower for

0.8 emissivity in comparison to the model without radiation heat transport. Even for the lower emissivity of the gas flow limiting areas and lack of absorption and scattering mechanisms within the fluid domain the radiation cannot be neglected.

5. Conclusions

The paper presents the new, 3D reduced model of the close coupled showerhead MOCVD reactor. The comparison of the results obtained with the aid of the new solution and the 3D model based on the symmetry of the chamber are in good conformity with each other. It has been proven that the novel approach incorporating only single line of nozzles allowed analysing phenomena taking place within the modelled chamber. Moreover, the importance of boundary conditions regarding proper modelling of the showerhead cooling are emphasised. Either constant temperature boundary condition or convective boundary condition must be used. Thermal radiation is a very important heat transport mechanism within MOCVD chamber where the huge temperature gradients between the showerhead and the susceptor are present.

6. References

[1] Pawlowski R P, Theodoropoulos C, Salinger A G, Mountziaris T J, Moffat H K, Shadid J N and Thrush E J 2000 *J. Crystal Growth* **221** 622–8
[2] Parikh R P and Adomaitis R A 2006 *J. Crystal Growth* **286** 259–78
[3] Theodoropoulos C, Mountziaris T J, Moffat H K and Han J 2000 *J. Crystal Growth* **217** 65–81
[4] Watwe R M, Dumesic J A and Kuech T F 2000 *J. Crystal Growth* **221** 751–7
[5] Waag A, Wang X, Fundling S, Ledig J, Erenburg M, Neumann R, Suleiman M A, Merzsch S, Wei J, Li S, Wehmann H H, Bergbauer W, Strassburg M, Trampert A, Jahn U and Riechert H 2011 *Phys. Status Solidi* C8 **7–8** 2296–301
[6] Li S, Wang X, Mohajerani M S, Fundling S, Erenburg M, Wei J, Wehmann H H, Waag A, Mandl M, Bergbauer W and Strassburg M 2013 *J. Crystal Growth* **364** 149–54
[7] Wang X, Li S, Mohajerani M S, Ledig J, Wehmann H H, Mandl M, Strassburg M, Steegmuller U, Jahn U , Lahnemann J, Riechert H, Griffiths I, Cherns D and Waag A 2013 *Cryst. Growth Des.* **13** 3475–80
[8] Coltrin M E and Mitchell C C 2003 *J. Crystal Growth* **254** 35–45
[9] Fotiadis D I, Kieda S and Jensen K F 1990 *J. Crystal Growth* **102** 441–70
[10] Zhong S Q, Ren X M, Huang Y Q, Wang Q and Huang H 2010 *Chinese Sci. Bull.* **55** 560–6
[11] Mitrovic B, Gurary A and Kadinski L 2006 *J. Crystal Growth* **287** 656–63
[12] Mitrovic B and Gurary A 2007 *J. Crystal Growth* **303** 323–9
[13] Breiland WG, Coltrin M E, Creighton J R, Hou H Q, Moffat H K and Tsao J Y 1999 *Mater. Sci. Eng.* **R24** 241–74
[14] Jensen K F 1994 Transport phenomena in vapour phase epitaxy reactors *Handbook of crystal growth (Thin film epitaxy vol 3)* (Amsterdam: Elsevier) chapter 13 pp 543-99

Acknowledgments
This work has been financially supported by the EU projects "GECCO" (Project No. 280694).

Verification of Thermo-Fluidic CVD Reactor Model

Z Lisik, M Turczynski, L Ruta[1] and E Raj
Lodz University of Technology, Department of Semiconductor and Optoelectronic Devices, Wolczanska 211/215, 90-924 Lodz, Poland

[1]E-mail: lukasz.ruta@p.lodz.pl

Abstract. Presented paper describes the numerical model of CVD (Chemical Vapour Deposition) reactor created in ANSYS CFX, whose main purpose is the evaluation of numerical approaches used to modelling of heat and mass transfer inside the reactor chamber. Verification of the worked out CVD model has been conducted with measurements under various thermal, pressure and gas flow rate conditions. Good agreement between experimental and numerical results confirms correctness of the elaborated model.

1. Introduction

Nowadays, the rapid development is observed in the area of semiconductor device manufacturing. Among the main technological steps, the processes oriented on thin-films deposition or growth of epitaxial layers are of crucial and still growing importance for microelectronics. Chemical vapour deposition (CVD) is one of them. Traditional CVD methods require high temperatures of 900-1000°C (or even higher) demanded for the activation of chemical reactions leading to the decomposition of gaseous reactants called precursors. As the result, the particles (atoms) necessary for deposited layer are released and the new layer grows on the surface of the heated substrate. The CVD technology is applied in many areas of industry when the epitaxial grow or thin layer coating surface are required, e.g. in production of protective coating on glass and metal, active and dielectric layers for semiconductor devices or multilayer structures for LEDs and lasers applications [1].

High temperatures required to thermal decomposition of precursors is, however, a serious limitation of traditional CVD in some processes e.g. limiting the number of materials that can be used as substrates. To reduce the process temperature to 400-500°C and omit this disadvantage, additional methods that allow enhancing the chemical reaction rates in lower temperatures are necessary. It can be achieved using the electrical activation of gaseous environment with the aid of plasma, like in the Plasma Enhanced Chemical Vapour Deposition (PECVD) technology. This is, presently, the most popular technology of low temperature CVD deposition [2].

The proper run of CVD processes is very sensitive to such factors as the state of gaseous environment or temperature field inside the reactor chamber. These parameters should be controlled and monitored during the reactor operation. Unfortunately, usually the reactor design does not allow or reduces possibility of full monitoring of the conditions inside the chambers during the deposition processes. For example, in the case of silicon carbide annealing processes that demand high and uniform temperature distribution inside the processing area, opaque crucibles are used that make impossible even the pyrometric temperature measurements [3]. One of the way to manage the problem is to employ numerical simulations that allow designing of multi-dimensional models of the heat and mass flow in the chamber under different operating conditions. Such numerical investigations of the phenomena taking place in CVD reactors have been already carried out and published [4-7]. Unfortunately, it was difficult to verify the correctness of the numerical approaches used in the models by direct measurements in the working reactors. Therefore, it has been found that it would be desired to perform some investigation aimed at the evaluation of the usually used numerical approaches, in which the simulations would be compared with the experiments carried out in a specially adopted CVD reactor. Such an investigation dealing with the experimental evaluation of the heat and mass transfer modelling inside a CVD reactor has been carried out and are reported in the paper.

2. CVD reactor used for experiments

As the object of investigations, the CVD reactor (Texas Instrument with build-in feasibility of enhancing chemical reaction rates with the aid of plasma) was chosen, as a typical reactor used to process thin film growth, which we could adopted to the planned experiments. Its photo is shown in figure 1. The diameter of its chamber was 59,5 cm, the diameter of the heating plate with susceptor playing the role of bottom electrode was 56 cm. The installed heater allowed to get the maximum temperature in the reactor reaching 400°C The working gas is supplied from the bottom and it flows along the susceptor to the outlet placed in the centre of the chamber.

Figure 1. The photo of the CVD reactor.

The preparation of CVD reactor to the experiments dealing with the examination of numerical approaches covered the arrangement of the reactor to create appropriate conditions for the considered measurements. As the result, the setup allows to carry out the experiments with atmospheric or lower gas pressure in the chamber, for gas flows changing from 0 up to 1000 sccm and the heater temperature of up to 300°C controlled by additional thermocouple. The adaptation of the reactor chamber to monitoring the temperature of processing samples was the other necessary modification. Its main part consisted of a set of thermocouples with measurement setup that have been introduced into the reactor chamber. Beside the thermocouple measurement system, the IR measurement procedure was also considered and appropriate tests have been performed. The CVD chamber covers two default quartz glass windows installed in the chamber walls, on the top and in the side wall, as it is shown in figure 2. In the first case, the window is placed directly above a small hole in the upper electrode, which allows monitoring the limited area of heating plate e.g. with the tested sample, whereas in the second case, the side window assures much wider field of view. In order to allow IR monitoring via the windows, their quartz-glass panes, having low transparency in the IR range, has been replaced the germanium ones.

Journal of Physics: Conference Series **494** (2014) 012020 doi:10.1088/1742-6596/494/1/012020

(a) (b)

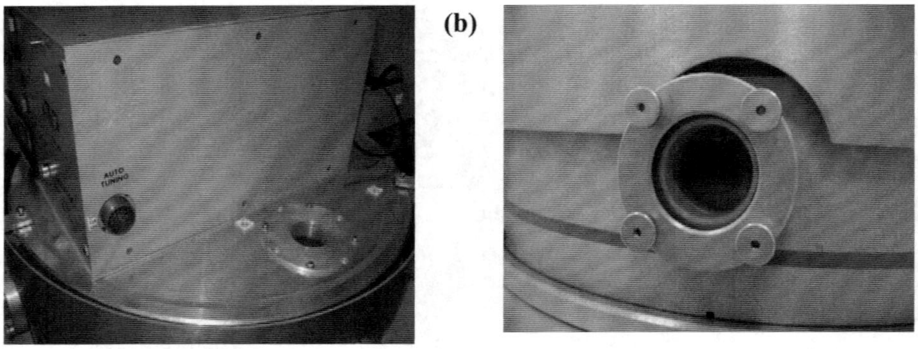

Figure 2. Windows in CVD reactor: (a) top quartz window, (b) side wall window.

In the IR measurements, the thermovision camera JENOPTIK VARIOCAM HR640 was used. The examples of IR pictures of the heated susceptor with and without a silicon pellet on it, taken directly by IR camera, are presented in figure 3. As it can be noticed, the colours of the susceptor "flows" whereas the difference between silicon pellet and the rest of susceptor is drastically large although their real temperatures are rather similar. To large extent, it is connected with the changes in the local measurement conditions such as the differences in emissivity, as it takes place in the case of the susceptor and silicon pellet. To be more useful, the pictures were converted in numerical scaling process by a special software, which reduced these adverse effects giving more realistic presentation of the temperature distribution.

Figure 3. IR pictures of heated suceptor at atmospheric pressure taken directly by IR camera (a) without silicon pellet, (b) with silicon pellet, (c) with silicon pellet after the scaling process.

3. Model of heat and mass transfer in CVD reactor

The 3D numerical model of heat and mass transfer in CVD reactor shown in figure 2 has been developed using commercial software ANSYS CFX [8] and it is limited to the domain shown in figure 4. As in typical models of such reactors [9-11], it covers the area of reactor chamber limited by its walls, the electrodes marked in red and the inlet and outlet tubes for working gases. The dimensions of the modelled structure are depicted in figure 4b. Since the model was designed to verify the used numerical approaches by the comparison of computed and measured results obtained for the same conditions, it imposed the basic assumptions for the carried out simulations that were as follows:

- Nitrogen is used as a working gas and its properties changes in agreement with the ideal gas law.
- Gas flow rate is constant and varies among simulations in the range 0 ÷ 1000 sccm.
- Reference pressure inside the chamber is in the range 7,6 mbar - 1 bar.
- Relative pressure at the outlet equals to 0 bar.
- Thermal boundary conditions on the wall surfaces are changing, being the subject of investigations.

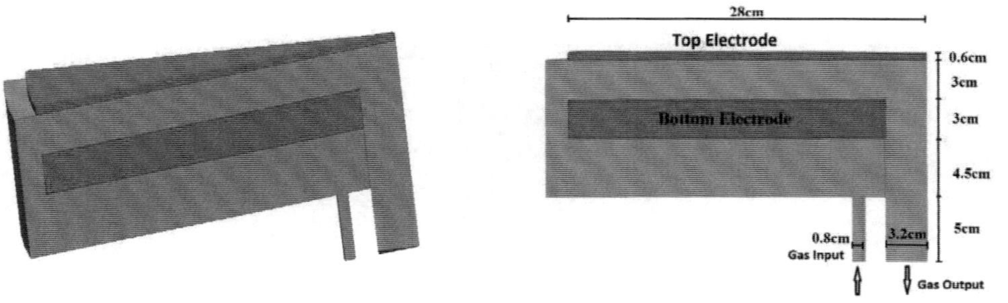

Figure 4. The geometry outline for the PECVD reactor chamber 3D model (a) spatial view of the model domain covering 15°-wedge chamber part, (b) radial cross section.

In the modeled CVD reactor, the heater in a form of a heating coil coated by the isolating ceramics is built into the bottom electrode that plays the role of susceptor. It means that the heat dissipation inside the bottom electrode, generally, is not homogeneous, which can lead to some inhomogeneity in the temperature distribution on the electrode surface influencing the gas-susceptor heat exchange. To verify such a possibility, the test measurements of the temperature distribution have been done using IR and thermocouple methods.

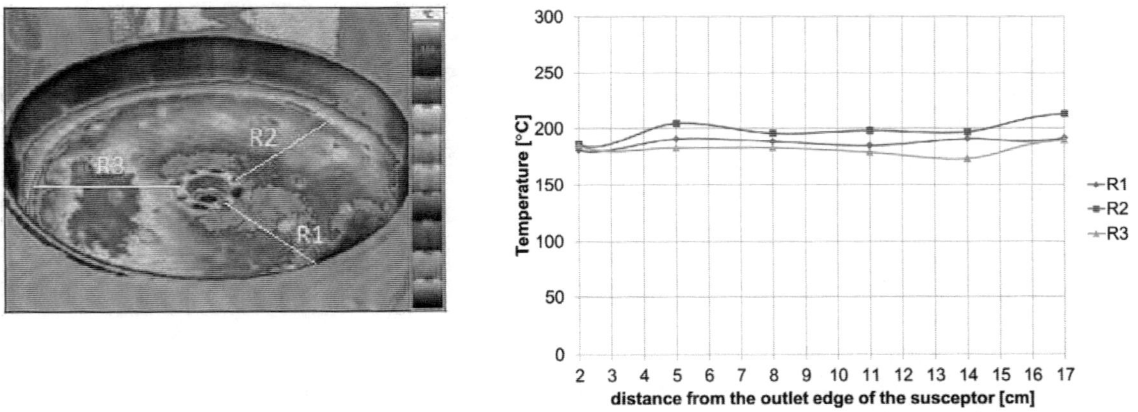

Figure 5. Temperature fluctuation on the susceptor surface represented by: IR measurements (left) and thermocouple measurements along three chosen radiuses (right).

The representative results are shown in figure 5. The IR measurements gave after the scaling process the color distribution like this one shown in figure 5a with an area of larger temperature observed always at the same place. The difference is larger than 16°C, which cannot be treated as reliable taking into account the uncertainty of the used IR temperature measurement method. Therefore, they have been followed by thermocouple measurements along the radiuses marked in figure 5a. The temperature deviations along the chosen radiuses are shown in figure 5b. They confirmed that no significant differences between the temperature distributions along the radiuses occurred and the "warmer" area indicated in figure 5a represents rather the fluctuation of measurement conditions only.

Nevertheless in figure 5b, some variations in the temperature measured along the radiuses are noticeable with a magnitude about 5 ÷ 8°C in relation to its average value. They look like stochastic ones without any relation to the heating coil location on the one hand and their magnitude is very small of the order of a few percents, on the other hand. It allows us representing the heater as a homogeneous layer located in the middle of the bottom electrode, as it is shown in figure 6. Its temperature T_S has been assumed as a constant one varying among simulations in the range 0 ÷ 300°C.

Journal of Physics: Conference Series **494** (2014) 012020 doi:10.1088/1742-6596/494/1/012020

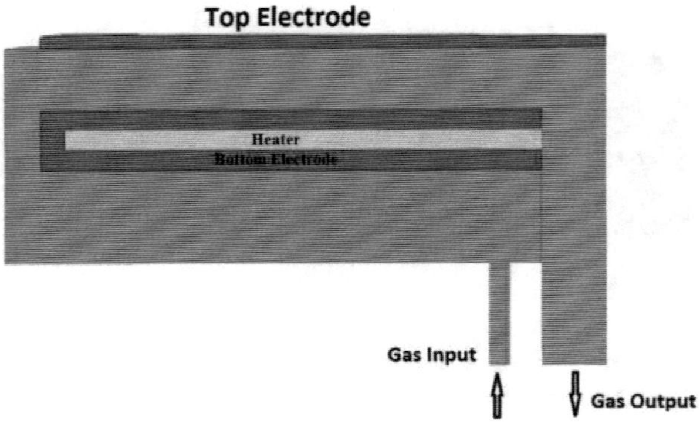

Figure 6. Sketch of the model domain presenting localization of the heater inside bottom electrode.

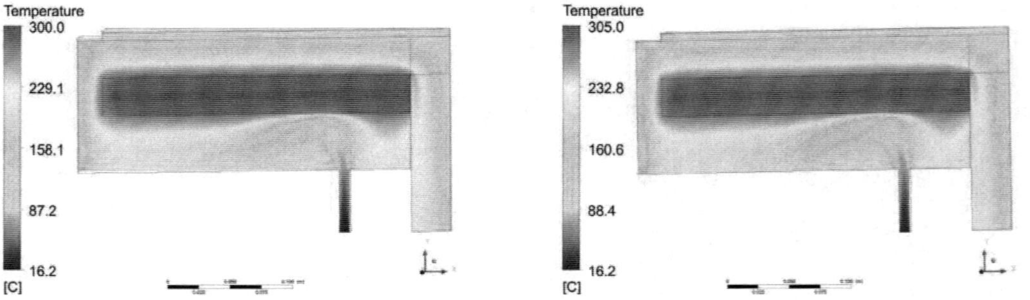

Figure 7. Temperature distributions obtained for: (a) homogeneous heater temperature distribution (b) sinewave heater temperature distribution ($T_{Saverage}$ = 300°C, sinewave magnitude ± 5°C).

In order to check if the presence of some disturbances in the heater temperature can affect the temperature distribution inside the reactor chamber working area, the comparative simulations for the same operating conditions and the average heater temperature but with homogeneous and sinewave distribution along the heater have been conducted. The obtained temperature distributions shown in figure 7 are almost identical, which confirms the correctness of proposed assumption.

4. Results of investigation

Using the model described above, series of simulations have been carried out changing the temperature enforced at the bottom electrode, the pressure and the gas flow inside the chamber. Moreover the heat exchange with the surrounding has been modelled with the aid of different boundary conditions at the walls of the chamber. In the case of adiabatic boundary condition, no heat transfer occurs, what means that the dissipated heat is removed by the working gas only. The isothermal and convective boundary conditions correspond to different intensity of heat exchange with the surrounding determined by the temperature of cooling system implemented at the chamber walls.
Moreover, the influence of thermal radiation on the total heat transfer within the reactor chamber has been investigated. All the simulations were followed by the comparative measurements using the built-in thermocouple measurement system. As the basic reference, the temperatures T_1 and T_2 located at the surface of the heating plate as it is shown in figure 8, were used in majority of experiments.

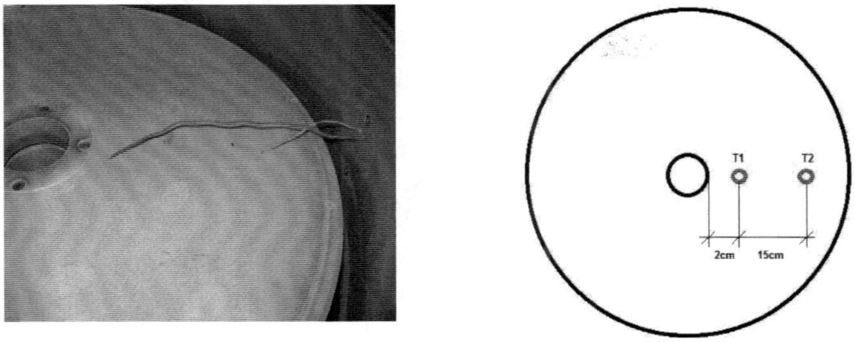

Figure 8. Localization of thermocouples in the chamber: general view (left) and the definition of two check points (right).

Series of simulations and experiments have been performed for different sets of parameters determining the operating conditions giving the full insight into the heat and mass transport processes in the reactor chamber. Below, some representative results are presented and discussed.

Figure 9 shows the simulation results covering the distributions of temperature and gas streamlines in the reactor chamber obtained for the adiabatic boundary condition on the surfaces limiting the modeled structure and the presence of radiative heat exchange inside the chamber with emissivity set to 0.2. The analysis has been carried out for the gas flow rate 1000 sccm, the pressure inside the chamber 1 bar and the heater temperature of 300°C. The temperatures measured in experiment were essentially lower than the one, obtained in the simulations. For example, the measured temperature T_1 on the susceptor was 275°C, whereas the value T_1 obtained in the simulation was 296°C. This difference may indicate that the heat removal by the gas flow is not the dominant process in the investigated reactor chamber.

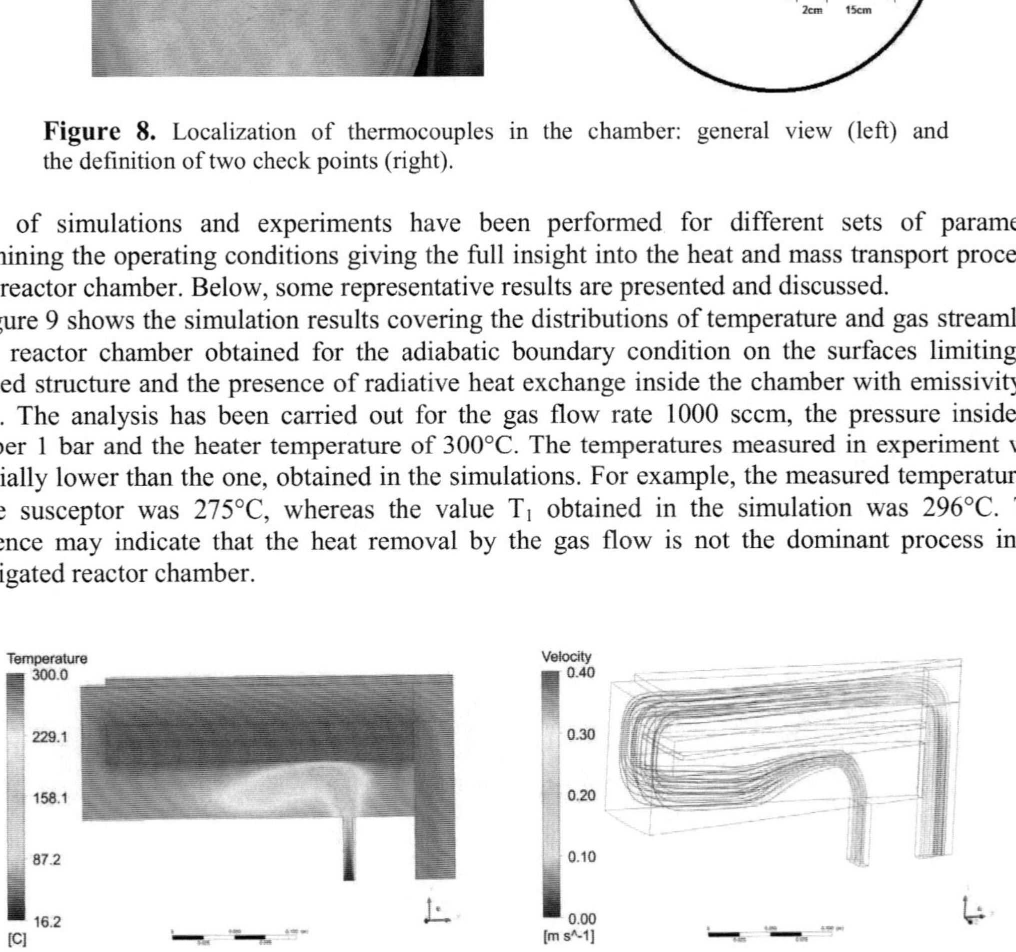

Figure 9. Results obtained for the adiabatic boundary condition: temperature distribution (left) and gasous streamlines (right) inside the CVD reactor.

Figure 10 shows the simulation results covering the distributions of temperature and gas pressure in the reactor chamber obtained for the convective boundary condition on the surfaces limiting the modeled structure with a typical values of the heat transfer coefficient of 10 W/cm^2K and the reference temperature of 25°C as well as the presence of radiative heat exchange inside the chamber with emissivity set to 0.2. The simulations performed with the convective boundary condition gave essentially different temperature distributions than previously. Now, the calculated value of the temperature T_1 on the susceptor is equal to 288°C and is much closer to the measured one. It confirms the suspicion that in the case of thermal processes in considered CVD reactor, the heat mass transfer connected with the gas flow is not the dominant path of heat transfer and the contribution of the heat exchange through the chamber walls must be taken into account.

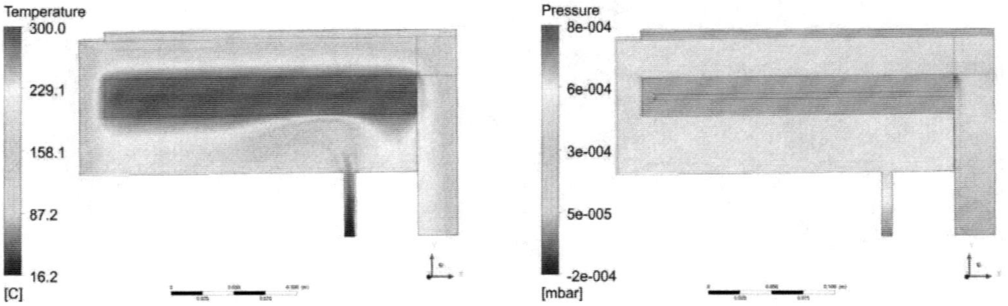

Figure 10. Results obtained for the convective boundary condition: temperature distribution (left) and gas relative pressure distribution (right) inside CVD reactor.

Taking the advantage of the CVD reactor measurement possibilities, an additional experiment has been carried out to confirm the correctness of the convective boundary condition. The temperature at the chosen point on the reactor side wall has been measurement using the thermocouple set-up and compared with the temperature at the same point evaluated numerically using adiabatic and convective boundary conditions. The comparison is shown in figure 11, which covers two curves representing the temperature distributions along the line covering the chosen check point (distance = 0 mm) taken from the 2-D distributions in figure 9 and figure 10, respectively. The black circle represents the thermocouple measurement. The measured value is about 10°C larger than that one evaluated numerically with the convective boundary condition being very far from the value estimated using the adiabatic boundary condition.

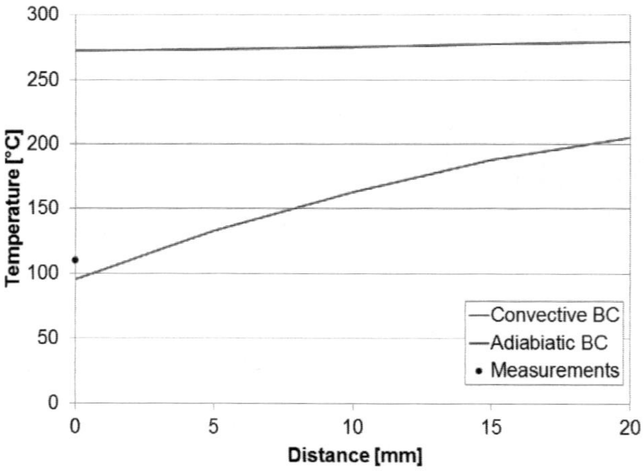

Figure 11. Temperature distributions at the wall of reactor chamber evaluated by the use of adiabatic and convective boundary condition and the temperature measured on the wall under the same operating conditions.

From the practical point of view, the main goal of any simulation is to evaluate the temperature in the processing area, i.e. in the susceptor area of CVD reactor. Therefore, the simulation results have been additionally compared with the temperature measurements along the line perpendicular to the susceptor plane, which begins at the check point T_1. In figure 12, the measured values are compared with the curves of temperature distribution extracted from figure 9 and figure 10 and the curves extracted from the simulations carried out for the same conditions but without the radiative heat exchange inside the chamber. The temperatures measured in the experiment were essentially lower than that one, obtained in the simulations not incorporating convection heat exchange with the surrounding and thermal radiation. The measured temperature T_1 on the susceptor surface was 281°C, whereas its value obtained in the simulation for the adiabatic boundary condition without radiation model was 299°C. This difference confirms that the heat removal by the gas flow is not the dominant process in the investigated reactor chamber also from the point of view of the evaluation of processing conditions.

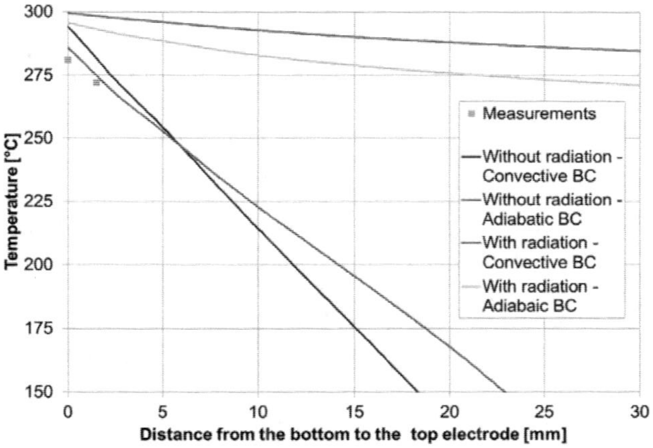

Figure 12. Temperature distribution along the line between electrodes located at point T1.

The results shown in figure 12 confirms that the heat removal from the CVD reactor chamber due to the convection phenomenon on the chamber walls is dominant but the participation of the radiation heat exchange can give also noticeable input to the final temperature distribution. It is rather small at the susceptor surface, and in the presented simulation it resulted in the difference of 8°C, but it increases in the area far from the susceptor, and at the top electrode the difference exceeds 45°C. This effect is clearly noticeable in figure 13, where the temperature distributions calculated for both the cases, with and without processes of radiative heat exchange, are presented.

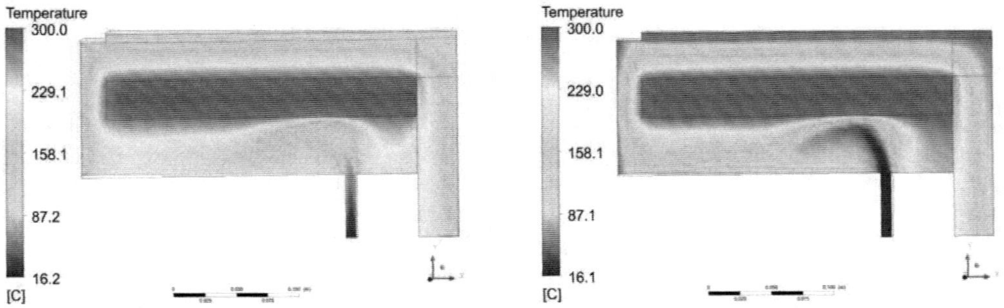

Figure 13. Temperature distributions in the CVD chamber obtained for the simulation with the radiative heat transport (left), and without the radiative heat transport (right).

The presented above results have been chosen and discussed as the representative one but they have been achieved for one set of parameters determining the work conditions i.e. the gas flow rate 1000 sccm, the pressure inside the chamber of 1 bar and the temperature of the heater 300°C. Their validity for the whole range of considered CVD reactor operating conditions has been confirmed also by many numerical and experimental investigations carried out for different set of these parameters.

Their results are gathered in a compact form in figure 14 and figure 15, for 1 bar and the lower pressure, respectively. The charts cover the comparison of temperatures in the check points T_1 and T_2 obtained experimentally and numerically using the convection boundary condition supported by the radiation heat exchange phenomenon. The difference between temperatures T_1 and T_2 in the numerical simulations is negligible, so only the single line has been plotted in the graphs.

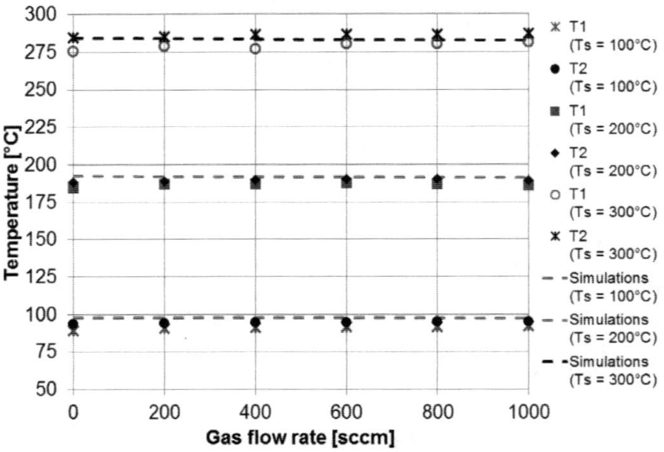

Figure 14. Comparison of simulations and measurement of the temperature dependence on the gas flow rate for the pressure inside the chamber of 1 bar.

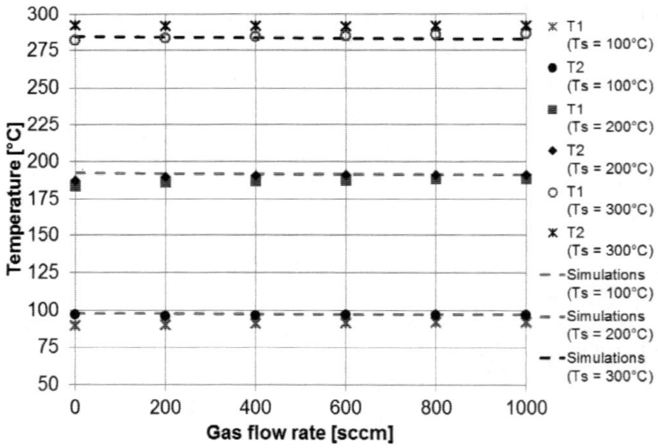

Figure 15. Comparison of simulations and measurement of the temperature dependence on the gas flow rate for the pressure inside the chamber of 7,6 mbar.

One can notice that the experimental points fit very well the curve obtained due to the numerical investigations in the whole range of considered operating parameters, which can be recognized as the prove of the correctness of used heat transfer models. In addition, the lack of noticeable influence of the gas flow velocity on the local temperature confirms the very small contribution of mass heat transfer in the investigated processes.

5. Conclusions

The paper deals with the evaluation of numerical approaches developed to create numerical models aimed at the simulation of the processes taking place in the CVD reactors chambers. Both experimental and numerical investigations have been employed with a specially adopted CVD reactor as the object of research. The rector has been adopted to work in the conditions typical for the film growth processes during semiconductors fabrication on the one hand and to create appropriate conditions for the necessary measurements, on the other hand.

The main goal of the carried out investigations was to develop and practically verify the numerical approaches allowing to design numerical models of heat and mass transfer taking place in the chambers of CVD reactors. Such numerical approaches has been created using ANSYS CFX commercial software and verified experimentally with the aid of the built-in thermocouple set up and IR measurements. The experiments proved that the developed numerical approaches allow credible modeling of the processes in CVD reactors. They indicated that the removal of heat dissipated in the susceptor heater takes place due to the convection heat exchange on the walls limited the reactor chamber with a little participation of the heat transfer connected with the gas flow. They also pointed out the role of the inner radiative heat exchange-in the total heat transfer inside the reactor chamber, which can result in significant changes in the temperature distribution far from the susceptor.

6. References

[1] 2009 *Chemical Vapour Deposition: Precursors, Processes and Applications* ed A C Jones and M L Hitchman (Cambridge: RSCPublishing)
[2] Beck R B 1991 *Technologia krzemowa* (Warszawa: WNT)
[3] Perek P, Raj E, Kubiak A 2009 Importance of convective transport mechanism in modelling of high-temperature reactor *Proc. Int. Conf. MicroTherm'2009* (Lodz, Poland, 25-28 June 2009) pp 313-8
[4] Beak S J, Kim Y J, Lee J C and Boo J H 2007 *J. Korean Physical Society* **51** 1113-8
[5] Cheng W T , Li H C and Huang C N 2008 *Chemical Eng. J.* **137** 603–13
[6] Mitrovic B, Parekh A, Ramer J, Merai V, Armour E A, Kadinski L and Gurary A 2006 *J. Crystal Growth* **289** 708–14
[7] Chen M, Yuan N, Ding J and Zhao Y 2012 *Lecture Notes in Information Technology* **9** pp 174-8
[8] 2013 *ANSYS CFX- Theory Guide* ANSYS Inc. rel. 14.5
[9] Nabati H, Mahmoudi J 2008 Fluid flow and heat transfer simulation in a barrel type CVD reactor *Proc. SIMS 2008 (Oslo, Norway, 7-8 October 2008)* www.scansims.org/sims2008 [accessed on July 2013]
[10] Saxena V, Muralidhar K, Eswaran V 2002 *SADHANA - Academy Proceedings in Engineering Sciences* **27 Part 6** 657–73
[11] Cheng T S, Hsiao M C 2008 *J. Crystal Growth* **310** 3097– 106

Acknowledgments
This work has been financially supported by the EU project "GECCO" (Project No. 280694).

AUTHOR INDEX

Institute of Physics
Dirac House, Temple Back
Bristol BS1 6BE UK

CURRAN ASSOCIATES INC
proceedings.com